This book, by a leading thinker with thirty years of experience in the field, is the first devoted to fibrous composites in biology. It tackles a major unsolved problem in developmental biology: How does chemistry create architecture outside cells? Fibrous composites occur in all skeletal systems, including plant cell walls, insect cuticles, moth eggshells, bone, and cornea. They function like industrial fibreglass, with fibres set in a matrix. The fibrous molecules are long, extracellular, and water-insoluble, and to be effective they must be oriented strategically. The underlying hypothesis of this book is that the fibres are oriented by self-assembly just outside the cells during a mobile liquid crystalline phase prior to stabilization. The most common orientations of the fibres are as plywood laminates (orthogonal and helicoidal) and as parallel fibres. These may be imitated *in vitro* by liquid crystalline chemicals.

The book takes an interdisciplinary approach and will be of interest to biologists, biochemists, biophysicists, materials scientists, and chemists working with liquid crystals.

Biology of fibrous composites

Biology of fibrous composites
Development beyond the cell membrane

A. C. NEVILLE
University of Bristol

UNIVERSITY PRESS

CAMBRIDGE
UNIVERSITY PRESS

32 Avenue of the Americas, New York NY 10013-2473, USA

Cambridge University Press is part of the University of Cambridge.

It furthers the University's mission by disseminating knowledge in the pursuit of education, learning and research at the highest international levels of excellence.

www.cambridge.org
Information on this title: www.cambridge.org/9780521410519

© Cambridge University Press 1993

This publication is in copyright. Subject to statutory exception and to the provisions of relevant collective licensing agreements, no reproduction of any part may take place without the written permission of Cambridge University Press.

First published 1993

A catalogue record for this publication is available from the British Library

Library of Congress Cataloguing in Publication data
Neville, Anthony C. (Anthony Charles), 1937–
Biology of fibrous composites : development beyond the cell membrane / A. C. Neville.
   p. cm.
Includes bibliographical references (p.   ) and index.
ISBN 0-521-41051-7 (hc)
1. Cytoskeleton – Formation.   2. Fibrous composites.
3. Developmental biology.   4. Cytoskeletal proteins.
5. Glycoproteins.   I. Title.
QH603.C96N48   1993
574.4'7–dc20   92-35618
                                                                 CIP

ISBN 978-0-521-41051-9 Hardback

Cambridge University Press has no responsibility for the persistence or accuracy of URLs for external or third-party internet websites referred to in this publication, and does not guarantee that any content on such websites is, or will remain, accurate or appropriate.

Contents

Acknowledgements — page **vii**

1. Defining the subject — **1**
   1.1. Introduction / 1
   1.2. Aims and significance / 5
   1.3. Definitions and diagrams / 6

2. The occurrence of fibrous composites — **25**
   2.1. Parallel fibres / 25
   2.2. Orthogonal plywoods and other large angular shifts / 27
   2.3. Helicoidal plywoods in animals / 39
   2.4. Helicoidal plywoods in plant cell walls / 64
   2.5. Pseudo-orthogonal plywoods / 77
   2.6. Anomalous distribution of helicoids / 81

3. Properties of natural plywoods — **85**
   3.1. Chemistry of fibrous composites / 85
   3.2. Mechanical functions: fibre orientation strategies / 89
   3.3. Optical properties / 105
   3.4. Influences of fibrous composites on cells / 116

4. Biomimicry: making liquid crystalline models of helicoids and other plywoods — **123**
   4.1. Principles and types of liquid crystals / 124
   4.2. Relevant properties established from synthetic materials / 125
   4.3. Natural liquid crystal models / 135

5. How is fibre orientation controlled? — **147**
   5.1. Self-assembly: hypotheses based on molecular shape / 148
   5.2. Directed assembly: hypotheses based on cellular mechanisms / 162
   5.3. Mechanical reorientation / 176

6. Unifying themes   **181**
   6.1. Interrelations and transitions between different architectures / 181
   6.2. Some general principles of helicoids / 185
   6.3. Evolutionary aspects / 191

References   **193**
Index   **211**

Acknowledgements

We all glean references cited in each other's papers, and I owe a huge debt of gratitude to numerous colleagues. This is especially important in an interdisciplinary field such as fibrous composites. In particular I would thank the following:

Firstly, for advice and information: Mr. N. Ashcroft, Professor E. D. T. Atkins, Professor P. Bonfante-Fasolo, Professor Y. Bouligand, Dr. D. H. Brown, Professor A. Buchala, Mrs. B. Costello, Dr. S. J. Hamodrakas, Dr. J. Kathirithamby, Dr. P. Marshall, Dr. S. E. Reynolds, Dr. B. Satiat-Jeunemaître, and Dr. B. D. Smith.

Secondly, for providing photographs of a wide variety of subjects: Dr. R. M. Abeysekera, Professor P. Bonfante-Fasolo, Miss W. Darke, Dr. J. D. Dodson, Dr. C. C. Doncaster, Dr. L. J. Gathercole, Dr. M. M. Giraud-Guille, Mr. J. Grierson, Dr. D. C. Gubb, Dr. S. J. Hamodrakas, Dr. P. M. Hughes, Dr. S. Hunt, Mr. M. Jackson, Dr. B. Kerry, Dr. L. Lepescheux, Dr. S. Levy, Mr. P. M. J. Loder, Mrs. B. M. Luke, the late Dr. J. N. Lythgoe, Dr. P. J. Miller, Dr. J. Newman, Professor J. Overton, Dr. A. J. Plumptre, Professor R. M. Rieger, Professor J.-C. Roland, Dr. B. Satiat-Jeunemaître, Sowerby Research Centre (British Aerospace, Bristol, U.K.), Professor W. H. Telfer, Dr. B. Vian, Professor M. Weidinger, Dr. J. H. M. Willison, and Dr. J. Woodhead-Galloway.

Thirdly, for provision of material: Dr. D. H. Brown, Professor J. D. Currey, Dr. B. Kerry, Mr. R. B. Kirby, Mrs. C. Lawson, Dr. C. Little, and Worldwide Butterflies (Sherbourne, U.K.).

Fourthly, I want to acknowledge expert technical assistance from Mr. T. Colborn (photography and technical drawing), Mr. A. Greagsby (photomicrography), Mrs. B. M. Luke (electron microscopy), Mr. S. Martin (electron microscopy), and Mr. M. Shannon (photography).

Finally, my special thanks to my wife, Monica, for putting up with the antisocial habit of authorship.

1
Defining the subject

1.1 Introduction

1.1a Fibres in development

The skeletal systems of animals and plants contain extracellular fibres. Throughout development these fibres are somehow manipulated into precise directions within the supporting tissues, so as to be best adapted to their mechanical and optical functions. How this is done constitutes the subject of this book. The problem, which is unsolved, is a facet of developmental biology which is usually neglected by authors of textbooks. This control of fibre orientation occurs in plants as well as animals: in plant cell walls (from algae to timber), in the cornea in birds and other vertebrates, in human bones, in basement lamellae, in arthropod exoskeletons, and in many other systems. It is therefore a truly *interdisciplinary* topic in biology, relevant to researchers in all of these fields and many more. Without accurate control of fibre architecture, many animals and plants would collapse.

Those macromolecules whose functions are skeletal, such as the polysaccharides cellulose and chitin and the protein collagen, are very long. They assemble in parallel to form microfibrils, fibrils, or fibres, depending upon the particular system involved. Their intrachain backbone bonds are covalent, whereas the lateral interactions between chains are by interchain hydrogen bonds. Such assemblies have a tendency to split in a fibrous manner, like wood, if pulled in the most vulnerable direction. This is because both the strength and stiffness of fibrous assemblies are about ten times weaker when stretched across rather than along the length of the molecular backbone. Furthermore, the continuous matrices which glue the individual fibrous components together are weaker still.

For these reasons, it is vitally important that skeletal fibres are *strategically oriented*, so as best to be able to cope with the mechanical stresses and strains acting upon them. In addition to forming fibres, skeletal molecules also form 'plywood' architecture, some of which resembles the plywood made from timber scraps and some of which is considerably more advanced. The main theme of this book concerns the morphogenesis of such plywood-like laminates in the supporting structures of both animals and plants.

Plywood assemblies are *extracellular*, and the fibrous molecules which form them are *insoluble* in water. The central problem addressed in this book may be formulated as a question: *How are large insoluble molecules manipulated precisely into position outside the cells which make them?* A general solution to this question is proposed. I suggest that the extracellular matrix which surrounds the fibrous molecules passes through a mobile phase during development. This idea involves two further concepts: (1) The matrix is envisaged as self-assembling, and (2) it is thought to pass through a *liquid crystalline* phase so as to form the appropriate plywood arrangement. The matrix bonds to the long fibrous molecules, moving them into position. The mobile matrix subsequently stiffens; whereas the *matrix* is thought to be the prime mover in morphogenesis, the patterns which are seen in the electron microscope derive from the *fibres* rather than the matrix.

One type of architecture (*helicoidal*, defined in Section 1.3) is like a universal plywood and is the most abundant kind of regular extracellular structure in living systems. It is found in nearly all kinds of animals and plants (Chapter 2). The principle of helicoidal structure is clearly very important in biology and is relevant to researchers in a wide variety of fields.

An attempt is made in Chapter 5 (Section 5.1) to explain helicoid formation in terms of molecular shape for the specific cases of hemicelluloses in plant cell walls, proteins in insect eggshells, and collagen in bone. This is aided by comparison with synthetic cellulose derivatives. In the future we may hope for a clearer understanding of extracellular morphogenesis in terms of the complex biochemistry of proteins and polysaccharides.

In some instances parallel layers of fibres are oriented at a *specific angle* with respect to some defined axis of the animal or plant, often with great accuracy (e.g. 3° in beetle cuticles). Here self-assembly cannot by itself provide an explanation of how this is done. The cells which make and secrete the fibrous molecules seem to have programmed and direct control of fibre orientation. There are several types of hypotheses for this category of *directed assembly*, and these are summarized in Chapter 5 (Section 5.2).

The two basic kinds of control are (1) *remote control* by extracellular self-assembly of a succession of fibre angle changes, giving rise to a helicoid and, (2) *direct cellular control* of precise fibre directions by directed assembly. It is important not to confuse these two concepts, as has sometimes occurred in the literature. Both of these are examples of *primary orientation*, which may be defined as orientation brought about by intermolecular forces operating between macromolecules in a mobile (liquid crystalline) medium.

There is a third kind of control defined as *secondary orientation* (Neville, 1967c). This is a *re*-orientation which brings about readjustment to an existing primary orientation. It is caused by some physiologically derived force. Examples are plant cell turgor pressure, muscular forces, blood pressure, growth, or some other mechanically derived force. The overriding effects of such mechanical forces cause changes to the original fibre orientations. If this happens to a regular system like a helicoid, the expected patterns may be predicted by calculation and displayed using a computer (Chapter 5, Section 5.3). The moderating effects of secondary reorientation on strength and stiffness can also be estimated.

1.1b Development beyond the cell membrane

There are many aspects to the study of developmental biology. The subject originated in classical descriptive embryology, usually taught with emphasis on the development of Amphioxus, frog, chick, and rabbit. Then the celebrated central dogma of molecular biology was established. This explained the genetic code of DNA, its transcription to messenger RNA (subsequently shown to be reversible), and its eventual translation to make proteins on the ribosomes. Some of these proteins function as enzymes which then control the chemistry of development. Gurdon (1974) has convincingly demonstrated that any nucleus from any somatic cell of an individual frog is competent to direct development into a whole frog when placed in the right cytoplasmic environment. This highlights the problem of differentiation: How do identical nuclei with identical DNA cause the cells of an individual to develop into different types? A further problem in development concerns the way in which patterns are established. These aspects are well summarized in various existing texts on developmental biology. They mostly concern events *inside* cells, together with cell–cell interactions. Yet there is also a whole range of developmental phenomena taking place *outside* of cells which is scarcely mentioned. This book is concerned with the developmental fate of the many important molecules which have been made either directly (as in the case of proteins) by the genetic code or indirectly (e.g. polysaccharides) by enzymes and then secreted through the outer cell membrane. The cell membrane arbitrarily determines what is included in this book. Thus the protein rods of the pellicle of *Euglena* lie just within the cell plasma membrane and hence are not included. The cellulose cell wall in such algae as *Valonia*, however, is outside the cell membrane and therefore is considered relevant.

The plasma membrane bounds the cell, separating cell contents (intracellular) from external products (extracellular). It consists of lipids, with a hydrophilic group at one end of each molecule and non-polar groups at the other end. The non-polar groups associate by weak van der Waals linkages to form a bi-layered sandwich. These links are labile (liquid crystalline), so as to permit breakage and subsequent resealing during passage of large quantities of substances into or out of a cell by pinocytosis or exocytosis, respectively. The non-polar interface has a very high electrical resistance (like that of paraffin wax, which has similar bonds), so that the membrane is not freely permeable to charged ions. The membrane also contains proteins which may protrude to either side of the lipid bi-layer. Some of these proteins are synthetic enzymes responsible, for example, for polymerization of monomers to form cellulose or chitin in appropriate types of cells. The significance of the cell membrane in extracellular developmental biology is that it represents the vital interface between the inside of the cell (where the nucleic acids have direct control) and the outside, where more remote control methods operate. Hence the sub-title of this book, and of this section.

1.1c *The export of proteins from a cell*

Proteins which are retained for use within the cell that made them (e.g. actin or myosin in muscle cells) are synthesized on ribosomes which lie loose in the cytoplasm. Those destined for export from a cell are synthesized on the ribosomes attached to the membranes of the rough endoplasmic reticulum (RER), and they are translocated across the membrane into its lumen (known as the cisternal space). They are packaged in vesicles which are transported to the Golgi apparatus (by membrane flow), eventually leaving the cell by exocytosis. It has been suggested that transport is effected by binding to microtubule proteins, moving along a microtubule in a conveyor-belt style, by assembly of the microtubule at one end and disassembly at the other.

A signal, consisting of a specific sequence of mainly hydrophobic amino acids, is recognized by a signal receptor present in the RER membrane. If it occupies a terminal position, this *signal sequence* (15 to 30 residues long) is cleaved from the rest of the protein during translocation across the membrane. In proteins with an internal signal sequence, such as chicken ovalbumin, translocation does not involve cleavage. The signal sequences of different organisms (e.g. bacteria, mice) bear strong similarities to each other, especially in their hydrophobic nature. This suggests that there is a general receptor protein in the RER membrane. This protein is not found in smooth endoplasmic reticulum, which is associated with synthesis of lipids for export. The presence of extensive RER in a cell indicates that it is secreting extracellular proteins.

It is therefore the signal sequence acting like a passport which selects those proteins to be exported, including those forming extracellular structures. Signal sequences are clearly of crucial importance in extracellular developmental biology. Genetic recombination experiments in bacteria have confirmed that the addition of a signal sequence to a non-secreted protein will lead to its export. By contrast, mutants in which an extracellular protein lacks a signal sequence retain such protein in the cell.

1.1d Previous reviews

Several works have been devoted to different aspects of fibre orientation control. General works have been cited in the influential chapter on extracellular materials by Picken (1962), in the review of helicoids by Bouligand (1972), and in a study of the asymmetrical array of fibres in otherwise bilaterally symmetrical animals (Neville, 1976). Some reviews are dedicated to fibre orientation in specific systems: chitin in insect cuticle (Neville, 1967c) and cellulose in plant cell walls (Preston, 1952, 1974, 1988; Frey-Wyssling, 1976). Other reviews concern cellulose orientation in wood tracheids (Mark, 1967; Boyd, 1985) and wood analysed as a fibrous composite (Jeronimidis, 1980). Several recent works are dedicated to the function of helically wound fibres in animals (Alexander, 1987, 1988; Wainwright, 1988). There is a chapter dealing with molecular aspects of extracellular materials in the book by Alberts et al. (1989). Molecular and mechanical functions are integrated in two books which are highly recommended reading: Wainwright et al. (1976) and Vincent (1982).

1.2 Aims and significance

The *academic* purpose of this book is to present a novel account of fibrous composites (materials which resemble fibreglass) in animals and plants. It aims to attract the interest of a wide range of workers, such as developmental biologists, biochemists, plant physiologists, animal physiologists, botanists and zoologists, liquid crystal chemists, biophysicists, and materials scientists. The objective is to promote new thinking and inspiration wherever fibrous structure is relevant to these fields, by enticing research workers to read about similar work on materials different from the ones they normally use.

The specific aims of the chapters are as follows. The first chapter includes definitions of the principles of fibres and fibrous composites, together with the architectures which they produce. Chapter 2 covers the distribution of various prominent types of fibrous architecture; this reveals the importance, in particular, of helicoidal plywoods. The accent in this book is on development rather than function; however, chemical, optical, and mechanical properties are discussed in Chapter 3. The fourth chapter concerns the probable involvement of liquid crystals in the formation of fibrous composites. In Chapter 5 we begin to see an integration of knowledge – chemical structure and shape, liquid crystalline self-assembly, and formation of fibrous composites.

The problem may be stated in question form: How does chemistry create architecture outside cells? This opens new avenues of thought, such as how the shapes of hemicelluloses in plant cell walls and of proteins in insect eggshells relate to helicoidal structure. It is exciting that the work with insect eggs provides a link to molecular genetics. In Chapter 6 I search for generalizations. Architecture, for instance, may override chemistry; two cathedrals can be architecturally the same and yet be built of different rocks, and two different cathedrals may be built from the same rock. An attempt is made to relate overall molecular shapes to fibrillar architecture and to interlink the main types of structures in animals and plants via types of liquid crystals.

In addition to academic aims there are also *applied motives* for studying fibres, some of which lead to funding of research. Detailed knowledge of plant fibrous systems helps in understanding the mechanical properties of commercially useful cell walls, such as flax, jute, hemp, and cotton in the rope and textile trades. The paper and gum industries are also based on plant fibrous products. Fibrous studies are involved in gaining a better understanding of the theoretical strength of timber, the method of breakdown of wood by fungi, and the functions of fibre in human diet.

There are also applied reasons for studying some of the animal fibrous systems (e.g. the silk industry). Important vertebrate fibrous systems include human bone, tendon, cornea, teeth, artery walls, and extracellular matrix. Several diseases are due to malfunction of the basement membrane (see Section 2.2f). Some insecticides work by altering the fibrous composite matrix in insect cuticle, while some weedkillers have similar effects on plant cell walls. For the future there is the prospect of trying to copy fibrous composites *in vitro* by biomimicry. And for geologists, whose interests extend to the past, extracellular fibrous structures are important because often they are the only parts which fossilize.

1.3 Definitions and diagrams

1.3a Layer deposition sequences

Many extracellular fibrous composites are secreted layer by layer in a sequence resulting in a laminate. Figure 1.1 shows some examples, with time's arrow indicated; this sometimes points in an unexpected direction, and mistakes occasionally occur in the literature. Diagrams are given in Figure 1.1 for plant cell walls, insect cuticle, vertebrate cornea, moth eggshell, and fish eggshell.

Invertebrates mostly secrete from a two-dimensional epithelial cell surface, so that the natural product is a laminated structure. The products of neighbouring cells are pooled (e.g. in insect cuticle) so as to avoid weakening junctions (Fig. 1.2C). The implication of laminated structure is that it may be broken down or

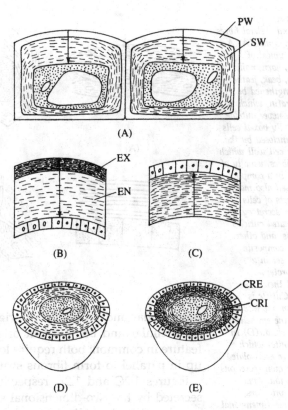

Figure 1.1. Simple diagrams of the sequence of deposition of some fibrous composites. Many extracellular supporting structures are laminates, secreted layer by layer in sequence (shown by the time arrows, which run from the older to the newer deposits). (A) Two neighbouring plant cells, surrounded by cell walls. The primary wall is secreted first (PW), and the secondary wall (SW) is secreted after the cell has finished enlarging. (B) Insect cuticle secreted by a cellular epithelium (epidermis). The exocuticle (EX) is secreted before an ecdysis and the subsequent expansion. This is followed by the endocuticle (EN), which is not expanded after secretion. (C) Cornea of the vertebrate eye. The epithelial cells on the outside of the eye secrete the primary stroma (shaded) via their inside faces. (D) A moth eggshell chorion (dashed) surrounds the oocyte (dotted), but is itself secreted by the maternal follicle cells of the female moth, which surround the shell. (E) A fish eggshell is also sheathed by the follicle cells of the female fish, but the shell layers are secreted from the inside by the oocyte; the first layer deposited is the cortex radiata externus (CRE), followed by the cortex radiata internus (CRI), which is helicoidal.

renewed only from its newest surface (e.g. during moulting in arthropods).

In vertebrates, by contrast, the secretion of fibrous composites is complicated by the invasion of cells. These cells become arranged in isolated suspension in three dimensions within the composite. Examples include fibroblasts (which secrete collagen) in the primary stroma of eye cornea, cells in sea-squirt tests (Fig. 2.29), and bone, which contains osteoclasts to break bone down, and osteoblasts to renew it. The shape of a bone can thus be continually remodelled from within. Although arthropod cuticle can change its thickness without a moult, moulting is needed to achieve changes in external shape.

1.3b Cellular and fibrous interrelations

Figure 1.2 shows some spatial relations between skeletal materials and the cells which secrete them. Figure 1.2A shows keratinocyte cells which form vertebrate structures such as horn, skin, feather, quill, and beak. These are not in fact examples of extracellular structures, because the fibrous protein keratin is found *inside* the cells. Plant fibres are built of individually boxed cells (Fig. 1.2B) (e.g. jute and hemp). Despite keratinocytes being

Figure 1.2. Simple drawings of interrelations of cells and skeletal fibres in some major living systems. (A) A group of keratinocytes typical of vertebrate tissues (skin, horn, nail, hoof, baleen, hair, claw, beak, feather, etc.). The cells are strengthened by fibres of the protein keratin, which is found inside the cells – never outside them. (B) The individually boxed cells of plants. Each cell is enclosed by the fibrous composite plant cell wall which it has itself secreted. Fibres, usually of cellulose, are embedded in a complex extracellular matrix, itself also mainly polysaccharide. (C) Sheets of cells (epithelia) combine their secretory efforts to produce laminated cuticles in insects, spiders, crabs, and other arthropods. The fibrous composite cuticle has chitin fibres set in a complicated matrix of proteins. Neighbouring cells pool their extracellular products. Cuticle is secreted exclusively from one face (the outer or apical face) of the epithelium; the cells are highly polarized. (D) In vertebrates, the cells (white) which secrete bone and cartilage are isolated by the products of their own secretions (stippled). They keep in touch via communicating cellular processes, running through the three-dimensional matrix. Unlike in arthropods, secretion is not confined to one localized part of the cell surface.

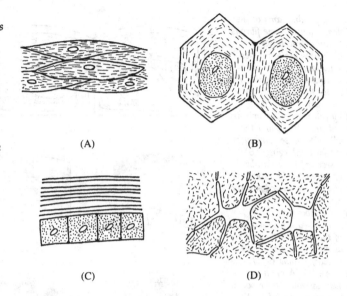

intracellular and zoological in origin, and plant cell walls being extracellular and botanical in origin, they nevertheless have a feature in common: both require long cells, with molecules lined up in parallel to form fibrous structures.

Figures 1.2C and 1.2D respectively show a pooled laminate secreted by the two-dimensional surface of an epithelium compared with a three-dimensional pooled matrix, as in bone and cartilage.

1.3c Microfibrillar crystallites as construction units

Different architectures are built with differently sized construction units (Neville, 1975b). In the case of cholesteric liquid crystals (see Section 1.3g and Chapter 4) of PBLG (polybenzyl-L-glutamate) (Fig. 4.7A), the construction unit is the single molecular chain. Biological fibrous composites are built with larger units, such as microfibrils or crystallites (e.g. cellulose in plant cell walls, or chitin in insect cuticle). These are of the order of 3 nm in diameter, in the case of chitin consisting of 19 molecular chains hydrogen-bonded in parallel. Details of the structure of a chitin chain and crystallite are given in Figure 1.3. Other composites are built of fibrils (bundles of microfibrils). Examples are crab cuticle (Fig. 2.23), beetle cuticle layers (Fig. 2.39), the cuticle in *Riftia pachyptila*, which belongs to the new phylum Vestimentifera (Gaill, Herbage, & Lepescheux, 1988), tunicate test cellulose (Fig. 1.14), and collagen in frog tadpole cornea (Fig. 2.8). In the case of crab cuticle the fibrils are 25–50 nm in diameter.

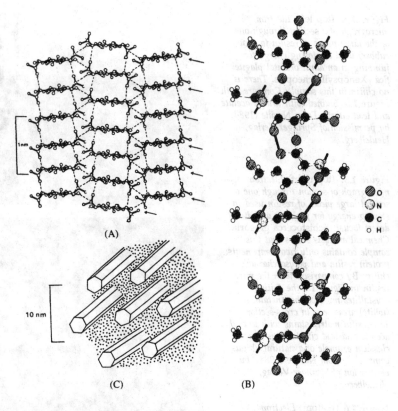

Figure 1.3. Chitin molecular and crystallite structure. (A) Chitin crystallites in arthropod cuticle are shaped like rods: this molecular model is viewed from the end of a rod. Nineteen molecular chains (see part B) make up the cross-section, arranged in stacks of 6, 7, and 6. Dotted lines indicate the strong hydrogen bonding between chains. The arrows indicate $n = 7$ refracting planes per crystallite; these give rise to $n - 2 = 5$ subsidiary maxima on the equator of the X-ray diffraction pattern (Prof. E. D. T. Atkins, personal communication). These small chitin rods form the equivalent of the glass in a fibreglass composite. Figure from Neville (1984), by permission of Springer-Verlag, Heidelberg, incorporating positions of hydrogen bonds as refined by Minke and Blackwell (1978). (B) Molecular model of a short length of a chitin chain (poly-N-acetylglucosamine). Types of atoms are shown in the key, and hydrogen bonds by solid rods; some of these hydrogen bonds stabilize the chain backbone, while others bridge across to overlying or underlying chains. From Neville (1978), © 1978, Carolina Biological Supply Co., Burlington, N.C., with permission; based on Carlstrøm (1957). (C) Diagram of seven chitin crystallite rods (see part A). The rods are shown in parallel, with hexagonal spacing (usually distorted to pseudo-hexagonal). The protein matrix is dotted. The rods are small and are drawn variously oriented about their own axes; they therefore give a fibre X-ray diffraction pattern rather than a more detailed single-crystal diffraction pattern. The chitin rods and protein matrix form a natural fibre composite. From Neville (1984), by permission of Springer-Verlag, Heidelberg.

1.3d Fibrous composites

A main theme of this book is fibrous composites, which function like fibreglass (Section 3.2). A good example is the rubber-like cuticle in insects, which contains crystallites of chitin embedded in a matrix of the protein resilin. This is a rubber with an almost perfect elastic efficiency (97% resilience). Because the crystallites are very small in diameter – 3 nm is much thinner than a cell membrane – how do we know that the crystallites are chitin and the matrix resilin? Figure 1.4 shows an electron micrograph of pure resilin from a locust; no crystallites are seen. Figure 1.5 shows a micrograph of a sample of rubber-like cuticle from a locust – known from chemical analysis to contain only chitin and resilin (Neville, 1963b). Hence, by comparing the two, we see that the crystallites resolved in Figure 1.5 must be chitin. More convincingly, the volume fractions of crystallites and matrix may be measured from the micrographs and multiplied by the respective dry densities of chitin and resilin. This gives good agreement with quantitative chemical analysis (Neville, 1984). The chitin remains unpenetrated by heavy-metal stains such as uranyl acetate or potassium permanganate; this indicates that the chitin is highly crystalline, and this is confirmed by the detailed X-ray diffraction diagrams obtained from tendon samples

10 / *Defining the subject*

Figure 1.4. (top left) Electron micrograph of a section through one of the large pieces of resilin (a protein rubber) used in storing energy for jumping in an adult oriental plague flea (Xenopsylla cheopsis. There is no chitin in this sample. Compare with Figure 1.5. Stained with uranyl acetate and lead citrate. From Neville (1984), by permission of Springer-Verlag, Heidelberg.

Figure 1.5. (top right) Electron micrograph of section through one of several large pieces of resilin used in storing energy for flight in an adult desert locust (Schistocerca gregaria). Chemical analysis shows that this sample contains only two components: protein resilin and polysaccharide chitin. By comparison with the pure resilin in Figure 1.4 the chitin crystallites must be the unstained (white) areas seen in cross-section. The resilin matrix stains with uranyl acetate and lead citrate. This is a classical example of a natural fibrous composite. From Neville (1984), by permission of Springer-Verlag, Heidelberg.

Figure 1.6 (bottom) Electron micrograph of an oblique section through the endocuticle of the fifth instar larval hind tibia of a giant water bug (Hydrocyrius colombiae; Belostomatidae). The architecture is pseudo-orthogonal and is visualized by chitin microfibrils as well as pore canal shapes (which, when crossing through parallel layers, appear like an aerial view of ships). Two major fibrous layers are seen; the change in direction between them is via approximately 90° of helicoid. Stained with potassium permanganate and lead citrate. Unpublished micrograph by B. M. Luke and A. C. Neville.

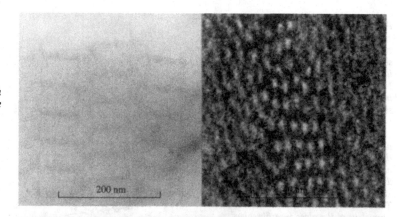

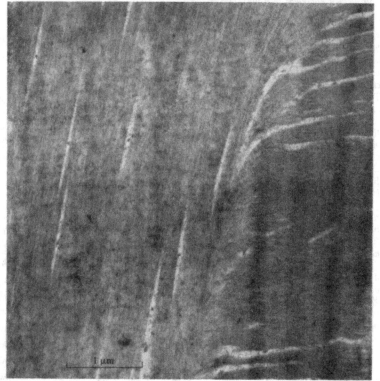

of insect cuticle (Rudall & Kenchington, 1973; Neville, 1975a). In order to function as a good fibrous composite, the fibrillar component must be as stiff (crystalline) as possible. For further details, see Section 3.2a and Rosen (1974).

1.3e Parallel fibre systems

Fibrous materials have very well oriented molecules which need to be *straight* to be able to line up in parallel (Fig. 1.6). In the case

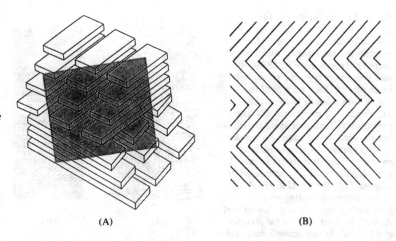

Figure 1.7. *Orthogonal 'plywood' and the pattern it produces.* (A) *Diagram to show two sets of layers of lath-shaped microfibrils oriented at right angles to each other. An obliquely inclined plane of section is shown by stippling; it relates to part B.* (B) *Diagrammatic representation of a section cut at an oblique angle to the surface of a series of layers arranged as in part A. The plane of section meets each layer of microfibrils at approximately 45°. This generates a herring-bone pattern and is seen in suitably angled sections through the materials listed in Table 2.1A. From Neville (1988c), by permission of Biopress Ltd., Bristol, U.K.*

of structural polysaccharides such as cellulose and chitin, the β(1 → 4) backbone linkages zig-zag alternately along the chain (Fig. 1.3B). The zigs cancel the zags, giving a straight line. The same is true of the alternating zigs and zags of the peptide bonds in β-pleated proteins such as silk fibroin. Starch is a polymer of glucose – just like cellulose – except that the backbone links do not alternately zig-zag; instead, they zig-zig-zig or zag-zag-zag! Hence the overall shape of the starch chain is a helix. In crystalline starch the helices lie in parallel; this is why iodine molecules (I-I) stain starch dichroically. The stain molecules fit down the insides of the helices to give an oriented effect (dichroism) as the material is rotated in a polarizing microscope. We may conclude that the important molecules in fibrous composites are straight.

1.3f Orthogonal plies and herring-bone patterns

If straight molecules – or bundles of them – are oriented in parallel to form a sheet, and such sheets are oriented mutually at right angles, then an orthogonal system results (Fig. 1.7A), as in industrial plywood. Orthogonal plies are common in biology (Section 2.2). Figure 1.7B shows how a section cut obliquely through an orthogonal plywood generates a herring-bone pattern.

1.3g Helicoidal plies and arced patterns

This is really like an orthogonal plywood (as in the preceding section), except that instead of an angle of 90° between neighbouring layers there is a much smaller angle (often between 10° and 20°). *Helicoids* are multidirectional plywoods; the term 'universal plywood' has been coined for them (Neville, 1975a).

Figure 1.8. Helicoids and the patterns they produce. (A) Diagram of a helicoidal stack of layers of microfibrils, generated by computer graphics (Levy & Neville, 1986). In each component layer the microfibrils lie in parallel. Each microfibril is drawn with a square cross-section. The direction of microfibrils changes progressively through a small angle, like the grain in the steps of a wooden spiral staircase. An oblique plane of section is stippled; it relates to part B. From Neville (1988), by permission of Biopress Ltd., Bristol, U.K. (B) A ranked pattern of nesting arcs generated by the oblique section in part A. Such patterns indicate the presence of a helicoid. In this example each part of a microfibril is represented by a single line. The pattern was generated by photographing a solid model built with glass and Letratone©. From Neville and Levy (1985), by permission of Cambridge University Press. (C) Another representation of a helicoid, in the form of a wedge-shaped model. Ranks of arcs form a typical pattern on the oblique face. The helicoid consists of planes of microfibrils, rotating progressively clockwise away from the observer. From Neville (1984), by permission of Springer-Verlag, Heidelberg.

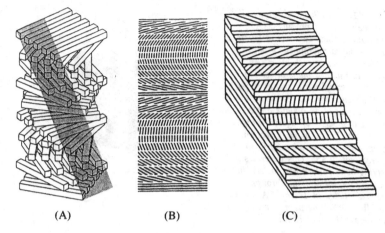

Each layer of crystallites is oriented in a parallel sheet. The direction of crystallites changes through a small angle from layer to layer in a constant sense of rotation. The spatial organization of the crystallites therefore resembles the grain directions in the steps of a wooden spiral staircase (Fig. 1.8A). An alternative description (conveying a dynamic sense of their formation) is that successive sheets of crystallites occupy the positions taken up by the wing of a sycamore seed as it rotates through the air.

Sections cut at an oblique angle to the planes of component layers of a helicoid generate characteristic ranks of arced patterning, as shown in Figure 1.8B. Such arced patterns enable us to detect the presence of helicoidal structure in electron micrographs. One set of arcs arises from each 180° rotation of layers (Fig. 1.8C). Each arc is composed of a number of straight segments which deceive the eye and make the observer think that the arcs are real rather than patterns. An example is shown in Figure 1.14. A computer-generated stereo pair of drawings of a helicoid is reproduced in Figure 1.9 (Levy, 1987). This may be viewed in three dimensions either by using a stereoscopic viewer or by means of a cross-eyed viewing technique described in the legend to Figure 5.7.

In the next chapter it will be established that helicoidal architecture predominates in animal and plant extracellular structures. Helices are the norm in biology, and the helicoid is really a form of discontinuous helix at a larger size level in the structural hierarchy. Helicoids form a major theme in this book. The history of their discovery is given in Sections 2.3 and 2.4. Their mechanical properties are considered in Section 3.2, optics in Section 3.3, chemistry in Section 3.1, liquid crystalline examples in Section 4.2, and morphogenesis in Section 5.1.

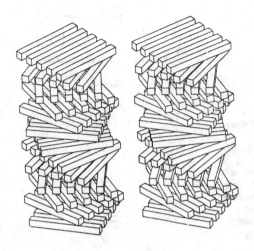

Figure 1.9. Helicoidal structure in stereo. A computer-generated stereo pair of diagrams of helicoidal structure. Component microfibrils are block-shaped and in natural examples would be set in a continuous matrix (not shown here). The microfibrils are mutually parallel in each plane, and there is a constant angular rotation from plane to plane. An oblique view through the 13 planes of microfibrils demonstrates the generation of arced patterning. These diagrams may be appreciated in 3-D by use of a stereoscopic viewer. Alternatively, a cross-eyed viewing method may be tried (see legend to Figure 5.7). From Levy and Neville (1986), by permission of Dr. B. Vian, Paris. A computer method for generating these diagrams is given by Levy (1986).

1.3h Microtomy artifacts in helicoids

Electron micrographs of sections cut obliquely through helicoidal materials show not only arced patterns but also light and dark banding for every 180° of rotation of the component layers. This gives the false impression that the chemistry changes every 180°; in reality the chemistry changes at a much finer level – between the periphery of each crystallite and the sheath of matrix in which it is embedded.

Bouligand (1986) gives a full explanation, including a mathematical appendix, of the dark and light bands which are superimposed on the arced pattern. He rejects the previous models of Gordon and Winfree (1978). In simple terms, the microtome knife used to cut sections for the transmission electron microscope meets the layers of crystallites at progressively changing angles. Some are met end-on, some sideways-on, and some obliquely. The knife therefore alternately cuts crystallites or passes between them; the cut ends are slightly turned up. This creates a rhythmical variation in section thickness (stepped model); thicker regions of a section will then appear more electron-dense than thinner regions. Bouligand's stepped model proposes that sections of helicoids have a *corrugated* surface, and this has been confirmed by Giraud-Guille (1986). Bouligand (1986) points out the important corollary of this topic. Taken at face value, an alternation of dense and less dense banding could *mistakenly* be interpreted as being the result of periodic changes in secretion by the relevant cells.

It is known that there are daily growth layers in insect cuticle (Neville, 1963a) and that these are due to an alternation in chitin crystallite orientation – from helicoidal (at night) to parallel (in the daytime); for further details see Section 5.2h. These daily layers can be seen in scanning electron micrographs of cut

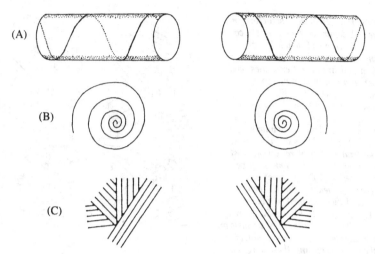

Figure 1.10. Differences among a helix, a spiral, and a helicoid; all three may exist in either left- or right-handed forms, which are mirror images of one another, as illustrated. (A) A helix is a coil which advances around a central axis. A diagonal straight line on a rectangular piece of paper forms a helix when the paper is rolled into a cylinder. (B) A spiral coils in a single plane. (C) A helicoid is a twisting 'plywood'.

surfaces of grasshopper cuticle (Banerjee, 1988a). A scanning electron micrograph which shows such layering in *Romalea microptera* is presented in Figure 5.20; it was taken by one of my students, who tackled the question of why such layering is visible using scanning electron microscopy (SEM) (Hughes, 1987). She found the cause to be the same as that giving rise to the layering discussed earlier for transmission electron micrographs of sections. The knife travels unevenly through the cuticular helicoid so as to create a corrugated surface, which is portrayed ideally by the scanning electron microscope. In higher-power micrographs the actual corrugations in helicoidal layers can be resolved by this technique. The parallel day layers do not show such corrugation, as expected; they act as a control.

While on the topic of artifacts, we may establish that arced patterning is not created by the preparatory techniques of transmission electron microscopy. Arcs may be seen by light microscopy in the large-pitch helicoid of sea-squirt test (see Section 2.3o) merely by focusing up and down in a cube of material, provided that this is tilted at an appropriate angle to the microscope axis.

1.3i Differences among helices, spirals, and helicoids

It is important to distinguish between these three systems, which are often confused or used unspecifically in the literature. All three may occur as mirror-imaged forms, with right or left rotation (Figs. 1.10 and 1.11). A *helix* (plural *helices*) is a coil which advances around a central axis. It may be modelled by drawing a straight line on a sheet of paper and rolling it into a cylinder (Fig. 1.10A). The French word for helix is *hélice*. Another term which relates (among others) to helices is *geodetic*. A geodesic is the shortest distance between two points on a curved surface; a

1.3 Definitions and diagrams / 15

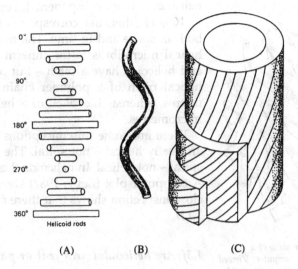

Figure 1.11. Diagrams to show the differences between helicoids and helices. Both may occur together. (A) Projection view of a sequence of straight microfibrils, rotating as an anti-clockwise helicoid. Rotation angles are shown. From Neville and Levy (1985), by permission of Cambridge University Press. (B) Drawing of an individual microfibril which follows a helical twist. (C) Diagram redrawn from Satiat-Jeunemaître (1989) showing a series of concentric cylindrical layers. Each layer is made up of a number of helices of uniform pitch. (If an individual layer is unrolled, the helices form a set of parallel lines.) The size of pitch of the helices in successive layers changes in a regular sequence. The whole sequence of concentric layers forms a cylindrical helicoid. The helicoidal layers in biological examples are usually deposited in a chronological sequence, so that the component layers form a succession in both space and time.

helix is therefore geodetic. Geodetic fibres are common in the cuticles of cylindrical animals. When such a skeletal element is distorted, the fibres begin immediately to take the strain; there is no slack to take up. Thus the full mechanical benefit of such fibres comes into play at once; this is why they are important.

A *spiral* coils (Fig. 1.10B) and expands at the same time – like a watch-spring. It may be either two- or three-dimensional. The term 'spiral' is often misused for 'helix'.

A *helicoid* (plural *helicoids*) is a system as defined in Section 1.3g, with successive planes of microfibrils. The French for helicoid is *hélicoïde*. A simplified sketch of a left and right pair of helicoids is given in Figure 1.10C. The pitch of a helicoid is the distance along its axis of rotation in which it turns through 360° (Fig. 1.11A). There are three ways to represent a helicoid in three dimensions: (1) by computer plot, to show a full turn through the 360° pitch of helicoid (Fig. 1.12) (Levy & Neville, 1986); (2) by a computer-generated stereo pair of diagrams with correct perspective (Levy & Neville, 1986); such a pair (Fig. 1.9) may be viewed in three dimensions either by stereoscopic viewer or by looking cross-eyed at the two images as described in the legend to Figure 5.7. (3) For an audience, a three-dimensional (3-D) effect may be achieved using two slide projectors set at the correct angle, one using a green image and the other a red image. The audience wear glasses with one red filter and one green filter, as used in early 3-D cinema for the horror movie *House of Wax*. Such glasses may be bulk-purchased. Suitable slides of helicoids have been made using computer plots and demonstrated in Bristol in 1986 by Sam Levy and Tim Colborn.

Helices and helicoids may co-exist in the same system. A diagram redrawn from Satiat-Jeunemaître (1989) illustrates a cylindrical plant cell wall cut away to show the relation between a

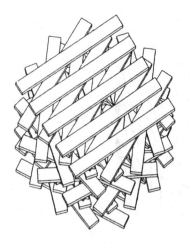

Figure 1.12. Surface view of a helicoid generated by computer. Viewed like this, arced patterns are not visible. This also shows another important point: that helicoids do not have just a single axis of rotation. Microfibrils are represented by plank shapes. They were drawn by a Calcomp plotter, driven from a GINO graphics package. The programme was executed on a Honeywell mainframe computer running the Multics operating system. A routine for hiding lines which are out of sight enables the image to be more readily understood. From Levy (1987), by permission of the author and Wissenschaftliche Verlagsgesellschaft, Stuttgart, Germany.

helicoid and its component layers of helical microfibrils (Fig. 1.11C). The helicoid corresponds to the succession of helices – both in space and in time. The most recently deposited layer of helical microfibrils is the innermost of the series. Both helices and helicoids have a pitch – but on different scales of size. The helical pitch of a polymer chain may be measured in nanometres, whereas the pitch of a helicoid is usually measured in micrometres.

There are some misconceptions in this field. Spiral staircases are really helical – not spiral. The genus of snails *Helix* is really spiral – not helical. In helicoids there is no single axis of rotation. A computer plot (Levy, 1987) shows a helicoid in surface view; close inspection shows that there are many axes of rotation (Fig. 1.12).

1.3j Are helicoidal arcs real or patterns?

In the seventies a controversy arose concerning the interpretation of arcs seen in electron micrographs – specifically those of arthropod cuticle. Dennell (1973, 1974) and some of his pupils (Mutvei, 1974; Dalingwater, 1975) believed that the arcs were real curved fibrils or microfibrils. We set out our arguments against this in detail in a paper on plant cell walls (Neville, Gubb, & Crawford, 1976). In favourable materials, arcs could be seen at high magnification as a pattern made up from straight microfibrillar segments. For instance, in electron micrographs of the test of a sea squirt (*Halocynthia*) the arcs are seen at low power (Fig. 1.13), but at higher magnification (Fig. 1.14) the arcs can be seen to be patterns or illusions made up of short elements. Each element is contributed from a microfibrillar sheet of the helicoid (Gubb, 1975). The same observation has been made on moth eggshells (Smith, Telfer, & Neville, 1971) and on the stonewort (*Nitella*) cell wall by Levy (1987).

Another powerful argument in support of helicoids is based on their unusual optical properties – namely, reflection of circularly polarized light and exceptionally high values of optical rotation which disperse with wavelength. This evidence is presented in Section 3.3b. The critical test to distinguish real arcs from arced patterns is based upon a tilting technique and is presented in Section 1.3k. Although the arced pattern is in effect an artifact, it can nevertheless be understood and used as a diagnostic indicator of helicoidal structure.

The Dennell controversy threatened the very existence of helicoidal interpretation, and yet the alternative did not lead anywhere. Helicoids provide a more productive hypothesis, leading to comparison with liquid crystals (Chapter 4) and to the beginning of an understanding of their origins in terms of molecular shape (Chapter 5).

Figure 1.13. Cellulose helicoid in an animal. Scanning electron micrograph of the face of an oblique section through the test of a Mediterranean sea squirt (Halocynthia papillosa; Tunicata). The architecture is helicoidal, and the component fibrils are made of cellulose (unusual in animals). The sense of rotation of components can be seen directly as anti-clockwise going into the material. From Gubb (1975), by permission of the author and Churchill Livingstone Journals, Edinburgh, U.K.

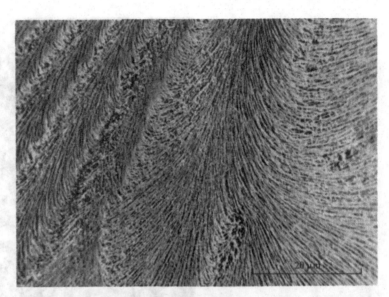

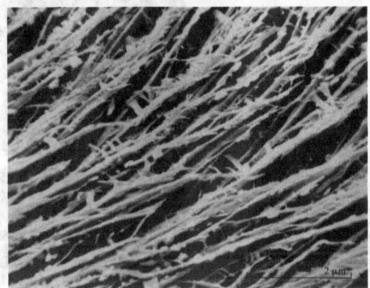

Figure 1.14. Proof that arcs are really patterns. As for Figure 1.13, but at higher magnification. Each arc (characteristic of an oblique section through a helicoid) is seen to be made up of components from several cellulose fibres; the arcs are therefore a pattern rather than a reality. From Gubb (1975), by permission of the author and Churchill Livingstone Journals, Edinburgh, U.K.

1.3k Critical tilt test for helicoids

Ranks of arced patterns are seen in electron microscope sections cut at an *oblique* angle through the layers of a helicoid, but are not seen in *vertical* sections of the same material. If, however, a vertical section is tilted about an axis running parallel to the helicoid layers – using a goniometric tilting stage – arced patterning then appears. What is more, the direction of the arcs reverses if the section is tilted the opposite way. An example of

18 / *Defining the subject*

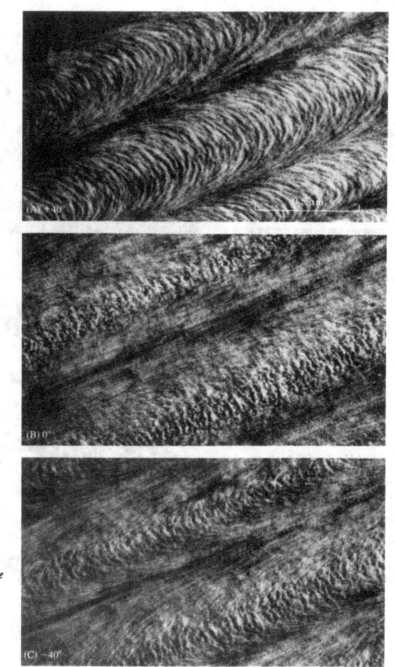

Figure 1.15. Proof that arcs are really patterns. Electron micrographs of a section cut obliquely through the spore wall of a fungus (Glomus epigaeum). The section is tilted through +40° in (A), untilted in (B), and tilted through −40° in the opposite direction in (C). This reverses the direction of the parabolic arcs (formed by chitin microfibrils), showing them to be patterns rather than actual curved fibres. From Bonfante-Fasolo and Vian (1984), by permission of the authors and Springer-Verlag, Wien.

this is shown for the fungal spore cell wall of *Glomus epigaeum* in Figure 1.15, from work by Bonfante-Fasolo and Vian (1984). Tilting also alters the apparent (but not the actual) pitch of the helicoid, as well as the shape of the arcs. By contrast, tilting does not affect the direction of arcs produced by canals (see Section 3.4a) in either arthropod cuticle or fish eggshells, because the crescentic shapes which they form when sectioned are *real* structures rather than patterns. This can be seen in micrographs of *Carcinus* cuticle (Livolant, Giraud, & Bouligand, 1978), where microfibrillar arced patterns reverse on tilting, but pore canal crescentic shapes in the same micrographs do not. The difference in response to tilting in microfibrillar versus pore canal systems can also be seen on the Perspex© model (see Fig. 1.18), as noted in our original legend (Neville & Luke, 1969b). Photographs of this tilted model showing arced pattern reversal are presented by Wharton (1978) and Neville (1980). The Perspex© helicoid is built from sheets ruled with parallel scratches to represent crystallites or microfibrils. The direction of scratches rotates helicoidally, and the sheets are then glued together. A face cut obliquely to the sheets then shows permanently an arced pattern. By contrast, a vertically cut face shows arced patterning only when tilted; the sense of the arcs reverses from positive to negative angles of viewing. A simpler way to build a model to demonstrate this principle is given by Livolant et al. (1978) using microscope slides on which parallel lines have been drawn at a progressively changing angle to the slide length. Alternatively, Letratone© may be used as in Figure 1.8B, with the slides placed in the slots of a storage box (Neville & Levy, 1985).

Dennell's curved arc alternative may be modelled very simply with a curved piece of plastic. Even when such a real curve is tilted about its axis of mirror symmetry through a full 360°, no reversal of arc direction is seen. Hence the Dennell model may be rejected, and the helicoidal model confirmed. This is the critical test for a helicoid.

Examples of tilting sections in the electron microscope to show arc reversal are seen in micrographs in the following literature: *Chara vulgaris* oospore wall (Neville et al., 1976); *Carcinus maenas* crab cuticle (Livolant et al., 1978); *Trichuris suis* nematode eggshell (Wharton, 1978); *Vigna radiata* mung bean cell walls (Vian et al., 1982); *Gadus morrhua* codfish eggshell (Grierson & Neville, 1981); *Nitella opaca* stonewort internode cell wall (Neville & Levy, 1984); *Homo sapiens* leg bone *helicoidal* osteons (Giraud-Guille, 1988). By contrast, sections of *orthogonal* bone osteons do not reverse their herring-bone pattern when tilted.

1.3l Left and right helicoids

Helicoids may be right- or left-handed, related to each other as mirror images (Fig. 1.16). The convention is to call a helicoid left-handed if the grain directions of its component plies turn anti-clockwise, about its axes of rotation, in layers successively *farther away* from the observer. The sense of rotation of specific systems is considered in Section 5.1e.

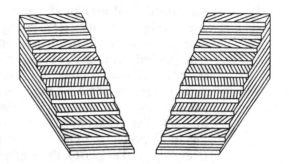

Figure 1.16. *The sense of rotation of helicoids determines the direction of arc pattern. Two wedge diagrams of helicoidal layers. Each oblique face generates ranks of arced patterns. The helicoid wedge on the left is anti-clockwise, with arcs pointing to the left; the helicoid on the right is clockwise, with arcs pointing to the right. So the direction of arcs can be used to determine the sense of rotation of a helicoid.*

1.3m Why are helicoids not detected even more often?

In the next chapter it will be established that helicoids are widely distributed in both animals and plants. We may therefore play devil's advocate and ask why helicoids are not detected even more frequently. The answers to this question involve the techniques for revealing helicoids, as previously considered (Neville & Levy, 1984).

(1) It is human nature to aim at vertical sections when cutting through a laminated material. But this will make a helicoid look like an orthogonal ply. Also, there is a tendency to reject any oblique sections, which if retained would have shown the arced patterning diagnostic of a helicoid.

(2) When investigating composite materials by electron microscopy, suitable staining is required to create contrast between crystallites and matrix. For instance, the arced pattern of locust cuticle is revealed by staining the matrix with potassium permanganate, but not with uranyl acetate (Neville & Luke, 1969b). Again, classical staining of sections of cell walls of the stonewort *Nitella* with uranyl acetate and lead citrate does not reveal arced patterning. Yet if the uranyl acetate is applied to the specimen while it is still in the unstained block of resin, the sections subsequently show arced patterns (Neville & Levy, 1984).

(3) Some plant cell walls require extraction prior to staining (e.g. with methylamine or dimethyl sulphoxide) in order to reveal arced patterning (Reis, 1981).

(4) Electron micrographs of helicoids taken at low magnification may show only the artifact banding coincident with 180° repeats. High power may reveal the characteristic ranks of arcs.

(5) Primary plant cell walls may show arced patterns only in sections taken at an early stage. Subsequent distortion during growth may *dissipate* the pattern (Roland et al., 1982; Pluymaekers, 1982). For this reason, arcs are more commonly seen in *secondary plant walls* which are deposited after growth in dimensions is complete – and which are not therefore subject

1.3 Definitions and diagrams / 21

to distortion. Arcs are also seen in the walls of spherical cells in which distortion forces are isodiametric (equal in all directions).

(6) It is possible to miss helicoidal structures using polarized light, if observation is confined to surface viewing. The helicoid fibrous directions will then mutually cancel to give a false isotropic (black) appearance.

(7) A similar result may occur with X-ray fibre diffraction. If investigation is restricted to aiming the beam in the same direction as a helicoid's axes of rotation, only a ring diagram will result. This gives the false impression of a random array of microfibrils. An example of this is given by Gaill and Bouligand (1987) for the helicoidally walled tubes of the deep-sea worm *Alvinella*. As they expected, an X-ray beam directed perpendicular to the surface revealed an apparent lack of preferred orientation of polymers, even though they had established the presence of helicoidal structure by electron microscopy.

1.3n Pseudo-orthogonal plies

This type of architecture appears superficially like a 90° plywood (orthogonal), but the angular change between major microfibrillar directions is gradual (via a quarter-turn of helicoid) rather than sudden. Hence our use of the term *pseudo* (= false) orthogonal ply. Pseudo-orthogonal structure was first encountered in insect cuticle (Neville & Luke, 1969b). A diagram of this architecture is given in Figure 1.17; it is drawn specifically for the example of wood tracheid cell walls, but applies equally well to the cuticle in some insects, including beetles and bugs. An electron micrograph of pseudo-orthogonal ply structure in a nymphal water bug (*Hydrocyrius colombiae*) is shown in Figure 1.6. Its distribution is discussed in Section 2.5, and its functional significance in Section 3.2f.

1.3o Two-system model of insect cuticle

We have analysed insect cuticles according to a two-system model with two alternating types of layers (Neville & Luke, 1969b). In one type the chitin microfibrils are oriented in parallel, often in line with some axis (e.g. wing or leg); this type is made in the daytime. The other type is helicoidal and is deposited at night. These two systems are shown on a Perspex© model (Fig. 1.18) and on a light micrograph taken in polarizing light (Fig. 1.19) of a leg section from a giant stick insect (*Heteropteryx dilatata*). At this magnification the helicoidal layers appear laminated. According to whether a particular sheet of microfibrils is seen end-on or longways-on, it will appear dark or brightly birefringent in the polarizing microscope. Microfibrils oriented at intermediate

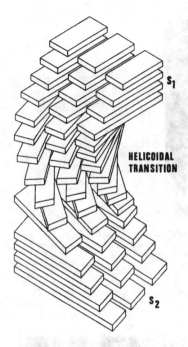

Figure 1.17. Pseudo-orthogonal structure. A diagram to interpret wood cell wall fine structure. From Neville and Levy (1985), by permission of Cambridge University Press. The three classical cellulose helices (Kerr & Bailey, 1934) of the secondary cell wall (S_1, S_2, S_3) spiral around the cylindrical cell, each at an angle characteristic for a given species. Only S_1 and S_2 are shown here. The S layers were long thought to cross each other at a large angle, but it is now established that the S layers are connected by thin intervening regions with helicoidal structure. We call such a combination of parallel and helicoidal layers pseudo-orthogonal, a term first used to describe the chitin directions in adult beetle cuticle (Zelazny & Neville, 1972). In wood, the building units are lath-shaped cellulose microfibrils. These lie in parallel in an S layer, and rotate anti-clockwise in the thin intervening helicoidal layers. The matrix is not shown.

22 / *Defining the subject*

Figure 1.18. Photograph of a Perspex© model for the two types of chitin architecture in insect cuticles. From Neville and Luke (1969b), by permission of Churchill Livingstone Journals, Edinburgh, U.K. Chitin microfibrils are represented by scratches ruled on a stack of Perspex© sheets. The lower part is of parallel microfibrils, typical of layers deposited during the day. The upper part is of helicoidally arranged microfibrils, typifying layers deposited at night. The lines on the sheets rotate progressively in an anti-clockwise direction. The sheets are glued together, and the model is sectioned obliquely at 45° to the laminated planes, generating arced patterning. Pore canals are cellular extensions which run from the epidermis through the cuticle to the outermost surface of the epicuticle. Holes representing pore canals are laterally compressed by the surrounding chitin fibrils into ribbon-like shapes. These twist as they traverse the helicoidal layers, but remain untwisted through the parallel day layers. The pore canals create crescents on the obliquely sloping face; these crescents are coincident with the arced pattern created by the scratches (microfibrils).

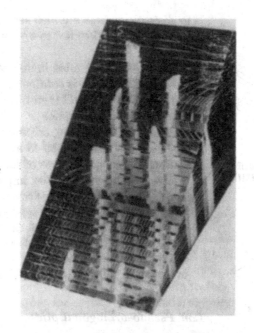

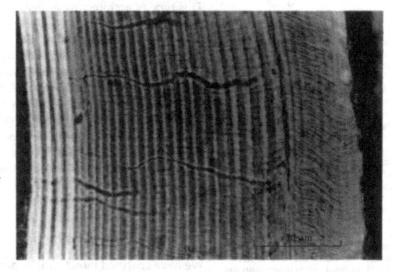

Figure 1.19. Polarizing micrograph of a section through the hind tibia of an adult giant Malaysian stick insect (Heteropteryx dilatata). The daily layer sequence is based upon the same principle shown in Figure 1.20A. By kind permission from Loder (1989).

angles will have intermediate brightnesses. Only in an electron microscope can the arced patterning be resolved, and this is confined to the helicoidal (night) layers. Hence a relationship can be established between polarizing and electron microscope images (Fig. 1.20). Once established, this relationship may be used to interpret the chitin microfibrillar architecture of various parts or appendages of the insect exoskeleton. The observations may be made in a polarizing microscope, but analysed as if an electron microscope had been used, thus avoiding the time-consuming protocol of the latter instrument.

Figure 1.20. Diagrams to correlate the appearance of daily layers of chitin architecture in insect cuticles, as seen by polarizing light microscopy (left) and electron microscopy (right). (A) Representation of a section of adult leg cuticle from a locust, killed 48 hours after the final moult between last-stage nymph (= larva) and adult. The exocuticle (EXO) is deposited before this moult. The daily layers are in the endocuticle. A layer (dotted) with chitin microfibrils parallel to the leg axis is deposited in the daytime (D1) following moulting; this is followed by a layer with helicoidal chitin microfibrils (N1), and so on. Because the chitin microfibrils of the day layers run in parallel and are perpendicular to the page, in cross-section they appear isotropic (dark) in polarized light. In the helicoidal regions a bright layer is seen wherever the chitin microfibrils lie in the plane of the page. Bright layers alternate with dark ones, where the chitin microfibrils in the helicoids rotate around to lie perpendicular to the page. The layers are deposited by the underlying epidermal cells (EPID). (B) This represents an electron microscope section cut obliquely to the leg axis. The helicoidal first night-time deposit (HEL) appears as ranks of arcs. In the example shown, the chitin directions rotate by 540°. The day layers are dotted to indicate chitin directions parallel to the leg axis (PAR). The electron micrograph is, of course, represented at a higher magnification than the light micrograph. Based on Neville (1984).

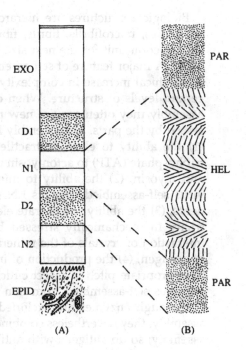

1.3p Self-assembly versus crystallization

Self-assembly is the spontaneous association of components to form a specific structure, whose design is determined solely by the bonding properties of the units. Because self-assembly features prominently in this book, it is relevant to compare and contrast it with crystallization. In thermodynamic terms, both proceed spontaneously towards a state of lower free energy, and hence towards greater structural stability. Both processes may be considered as if they occur at constant temperature. In general, crystallization is driven by enthalpy, by a decrease in the heat emitted; self-assembly is entropy-driven, by an increase in disorder, usually by exclusion of ordered water. As water bound onto component molecules detaches and loses order, the components self-assemble and increase order.

Neither templates nor enzymes are necessary for self-assembly; neither are they needed for crystallization. Self-assembly is perhaps a sophisticated form of crystallization, but it may involve limits to dimensions. Theoretically there is nothing to prevent a crystal of copper sulphate from growing a mile high – given suitable support. But a self-assembling system (e.g. myosin filament or collagen microfibril) may stop growing at a critical diameter; any further molecules added to the outermost layer would be unstable. Critical size may also be achieved, however, by crystallization (e.g. the diameters of a double helix of DNA or a region of α-helix in a protein).

Biological structures are hierarchical (e.g. molecules, macromolecules, microfibrils, fibrils, fibres, each of which forms the construction unit for the next size level in the hierarchy) (Crick, 1953). A major feature of self-assembly systems seems to be this hierarchical increase in complexity – *new properties* may arise at certain levels of structure. When components combine by self-assembly they often acquire new properties which are not possessed by the parts. The assembly is holistic. Some examples are (1) the ability to elicit contractile work by adding adenosine triphosphate (ATP) to actomyosin made by crudely mixing actin and myosin, (2) the ability to infect a host by tobacco mosaic virus self-assembled from its RNA and protein sheath components, (3) the ability to generate electric potentials by piezoelectricity in mechanically stressed bone, made possible by the interaction of crystals of the mineral hydroxyapatite with fibrils of collagen, (4) the production of iridescent colours by helicoids of appropriate pitch in insect cuticle, and (5) catalytic ability in enzymes self-assembled from non-catalytic sub-units.

Although enzymes are excluded from the definition of self-assembly, they nevertheless combine with their substrates by self-assembly; so do antigens with antibodies. Taking this approach to its logical conclusion, it could be argued that all physicochemical reactions proceed by self-assembly and that we are dealing merely with a conceptual field. Is 'self-assembly' just a euphonious and biological term for crystallization in inanimate systems? I think it is distinct: In particular, self-assembly may produce new properties and automatically limit dimensions. A further distinction is that fault disclinations are found only in liquid crystals and their biological analogues, not in solid crystals.

1.3q Self-assembly versus template-directed assembly

As we have seen in the preceding section, self-assembly is specified by the bonding capabilities of the component molecules. Template-directed assembly is different, as it involves agents in addition to the molecular components themselves. For example, microtubules are widely thought to play an important part in the positioning of cellulose microfibrils in plant cell walls. For a thought-provoking review comparing self-assembly with template-directed assembly at various levels in the structural hierarchy of plant cell walls, see Jarvis (1992). This and other examples are discussed in Section 5.2.

Having defined some principles and some types of architecture, we may next survey the distribution of fibrous composites in living systems.

2

The occurrence of fibrous composites

This chapter presents a wide range of examples of fibrous composites in animals and plants. A sensible amount of such 'stamp collecting' is justified because it shows the widespread distribution and importance of fibrous architecture (especially plywood types) in biological systems. The presentation in this chapter is systematic (by taxonomic groups) rather than comparative, for ease of access. The various types of material are described, illustrated, and listed in tables. Systems used in experiments on control of fibre directions are included.

2.1 Parallel fibres

2.1a Insect tendons and ovipositor drills

In examples where mechanical stresses are likely always to act upon a structure from a constant direction, the fibrous component can be aligned in parallel (as in Fig. 1.3C). This applies to the chitin crystallites in the tendons of insects, on which a muscle always pulls from a specific direction; it is well shown in the long tendon running throughout the length of the hind tibia to operate the tarsal joints in the large hind legs of crickets (e.g. *Decticus verrucivorus*) and grasshoppers (e.g. the locust *Schistocerca gregaria*). A scanning electron micrograph of a locust tendon showing parallel chitin is shown by Bennet-Clark (1976). Other good examples are the ovipositors used for drilling wood by wood wasps (e.g. *Sirex gigas*) (Rudall, 1967) and their ichneumonid parasitoid (*Rhyssa persuasoria*). These samples of parallel-oriented chitin are ideal for X-ray diffraction (Rudall, 1967) and for infra-red absorption studies (Neville, 1975a, 1980).

2.1b Daytime layers in insect cuticle

The orientation of chitin microfibrils in many insect endocuticles depends upon whether they were laid down during the daytime or the night-time (Fig. 5.21). Night cuticle has helicoidal architecture (Fig. 1.20), whereas that made during the day has parallel

26 / *The occurrence of fibrous composites*

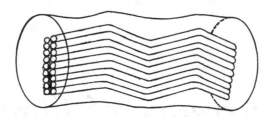

Figure 2.1. Diagram of crimping in the collagen fibrils of rat tails. A fibre containing several crimped fibrils is shown. These crimps permit extra extension of the fibre when strained. Redrawn from Nicholls et al. (1983).

chitin microfibrils; this is the basis of our two-system (Fig. 1.18) model for cuticle (Neville & Luke, 1969b). The direction of chitin in the day layers usually coincides with the axis of some major component, such as a leg femur or tibia, wing vein, or antenna. It functions to counteract bending stresses. These daily changes in chitin orientation are timed by a circadian clock (Neville, 1965b). In my early papers the day layers were called 'preferred' or 'non-lamellate' layers, and the night layers were called 'lamellate' layers. We now know from subsequent work (Neville & Luke, 1969a,b) and from the influence of Bouligand's interpretation of cuticular helicoids in crabs (Bouligand, 1965) that non-lamellate layers equate to parallel or unidirectional chitin, whereas lamellate layers are a manifestation of helicoidal layers (Fig. 1.20). In locusts (e.g. *Schistocerca gregaria* and *Locusta migratoria*) the chitin orientation can be experimentally manipulated by uncoupling the circadian clock (see Chapter 5) so as to produce endocuticle which has unidirectionally parallel-oriented chitin crystallites throughout.

2.1c Silk

Commercial silk, produced by the larvae of the silk moth (*Bombyx mori*, now extinct in the wild), is a good example of a fibre with molecules in parallel. The silk protein fibroin is extruded from the spinneret of the caterpillar in what is probably a nematic liquid crystalline state (see Chapter 4). Each cocoon is spun according to programme from a single thread, which is drawn into a finer thread ($\frac{1}{3}$ diameter) as it passes through the orifice. The long parallel silk molecules are already lined up in parallel in the liquid crystalline state; they are crystallized together by formation of hydrogen bonds between chains, resulting in a fibre of high tensile strength.

2.1d Vertebrate tendons

Like insect tendons, vertebrate tendons also consist of parallel, high-tensile-strength molecules – in this case of the striated protein collagen. This performs the same function as chitin in insect tendon. The discovery (Diamant et al., 1972) that collagen in vertebrate tendons exists in a crimped form (Fig. 2.1) has caused a basic reappraisal of collagen analysis by X-ray fibre diffraction. A favourite material for investigation of collagen is rat tail tendon; this turns out to be crimped. The crimping can be seen even in a light microscope using polarized light. (The moral of the story is that materials should be examined with low-power techniques before progressing to more prestigious technology.) The crimping has a planar waveform with the crimps in phase; it therefore gives different responses to compression forces according to which plane is involved (Nicholls et al., 1983). Under low tensile stresses, a tendon takes up the slack by uncrimping; only with larger stresses does the tendon have to take the strain directly on the collagen fibres themselves. In humans this device lessens the chance of damaging the Achilles tendon during strenuous activities. The crimps straighten and recrimp during non-injurious strain of the tendon. Collagen is also crimped in vertebrate skin and heart pericardium – anywhere in fact that it is subjected to tensile forces. There is, however, no crimping in bone collagen.

2.1e Plant cell walls

The guard cells of stomata possess cell walls with cellulose microfibrils oriented in parallel. Other cell walls may act as light guides or fibre optics (see Section 3.3f). Plants which are used as commercial fibres (ramie, flax, jute, hemp) also have predominantly parallel cellulose orientation in their fibre walls.

2.2 Orthogonal plywoods and other large angular shifts

An orthogonal plywood is built up from component layers which each contain one or several parallel planes of fibrils or microfibrils. There is a change in direction between neighbouring layers of 90° in strictly orthogonal plies (as in conventional industrial plywood), or of some other angle lying between 45° and 90°. A diagram of an orthogonal system is given in Figure 1.7, together with the herring-bone pattern generated by a section cut at a suitable angle through it. Orthogonal plies are abundant in biological systems (Table 2.1A), and some important examples are discussed later. They function in general as strong 'plywoods' and sometimes act to control shape by a lazy tongs effect (e.g. in the pig roundworm *Ascaris*).

2.2a Algal cell walls

An excellent example (Fig. 2.2) of a cell wall with orthogonally oriented cellulose microfibrils is seen in the alga *Eremosphaera viridis* (Weidinger & Ruppel, 1985). Orthogonal systems are common in algae, as shown in Table 2.1A, and include some of the most frequently studied algal cell walls (e.g. *Oocystis* is a

Table 2.1A. *Selected systems with sheets of parallel fibres oriented mutually either at 90° (orthogonal) or at some other large angle. Many systems occur in the body wall in so-called fibre-wound animals (Wainwright, 1988). Some systems previously regarded as orthogonal are now known to be pseudo-orthogonal (see Table 2.12).*

Group	Genus & species	Reference
Algae	*Oocystis solitaria*	Robinson & Herzog (1977)
	Eremosphaera viridis (cell walls)	Weidinger & Ruppel (1985)
	Chaetomorpha	Preston (1952)
	Apjohnia	Preston (1974)
	Cladophora	Preston (1974)
	Dictyosphaerica	Preston (1974)
	Glaucocystis	Willison & Brown (1978)
	Siphonocladus	Preston (1974)
Platyhelminthes		Clark & Cowey (1958)
Nematoda	*Ascaris lumbricoides* (cuticle)	Bird & Deutsch (1957)
		Harris & Crofton (1957)
Nematomorpha	*Paragordius varius*	Swanson (1974)
Nemertea	*Amphipora lactiflorens* (basement lamella)	Cowey (1952)
Annelida	*Chaetopterus* (dwelling tube)	Brown & McGee-Russell (1971)
	Lumbricus (cuticle)	Rudall (1968)
	Allolobophora (cuticle)	Reed & Rudall (1948)
	Paralvinella grasslei (cuticle)	Lepescheux (1988)
	Eisenia foetida (cuticle)	Richards (1984)
	Alvinella pompejana (cuticle)	Gaill et al. (1988)
Pogonophora	(integument)	Gupta & Little (1970)
Priapulida	*Priapulus caudatus* (basement lamella)	Morritz & Storch (1970)
Sipunculoidea	*Phascolion strombi* (basement lamella)	Morritz & Storch (1970)
Chaetognatha	*Sagitta elegans* (basal lamina)	Ahnelt (1984)
Arthropoda	*Oryctes rhinoceros* (eggshell chorion; orthogonal systems are very rare in arthropods)	Furneaux & Mackay (1976)
Mollusca	*Lolliguncula brevis* (mantle collagen tunic)	Ward & Wainwright (1972)
	Loligo pealei (mantle collagen tunic)	Ward & Wainwright (1972)
Chordata	*Bombinator* (tadpole) (epidermal basement lamella)	Rosin (1946)
	Xenopus laevis (tadpole) (epidermal basement lamella)	Overton (1979)
	Amblystoma opacum (larva) (epidermal basement lamella)	Weiss & Ferris (1956)
	Fundulus heteroclitus (basement lamella)	Nadol et al. (1969)
	Scyliorhinus caniculus (egg capsule)	Knight & Hunt (1974b)
	Latimeria chalumnae (scales)	Giraud et al. (1978)
	Amia calva (scales)	Meunier (1981)
	Pleuronectes platessa (scales)	Darke (1986)
	Pleuronectes limanda (scales)	Darke (1986)
	Pleuronectes microcephalus (scales)	Darke (1986)
	Hippoglossus vulgaris (scales)	Darke (1986)
	Rana temporaria (spectacle of eye)	Lythgoe (personal communication)
	Mus domesticus (cornea)	Haustein (1983)

Table 2.1B. *As for Table 2.1A, but for twisted orthogonal layers in bird corneas*

Species	Common name	Reference
Gallus domesticus	Domestic fowl	Coulombre & Coulombre (1975)
Columba livia	Pigeon	O'Donnell (1992), Poklewski-Koziell (1992)
Sturnus vulgaris	Starling	O'Donnell (1992), Poklewski-Koziell (1992)
Coturnix coturnix	Quail	O'Donnell (1992), Poklewski-Koziell (1992)
Perdix perdix	Partridge	O'Donnell (1992), Poklewski-Koziell (1992)
Phasianus colchicus	Pheasant	O'Donnell (1992), Poklewski-Koziell (1992)
Meleagris gallopavo	Turkey	O'Donnell (1992), Poklewski-Koziell (1992)
Oxyura jamaicensis	Ruddy duck	O'Donnell (1992), Poklewski-Koziell (1992)
Anas acuta	Pintail duck	O'Donnell (1992), Poklewski-Koziell (1992)
Anser caerulescens	Snow goose	O'Donnell (1992), Poklewski-Koziell (1992)
Branta leucopsis	Barnacle goose	O'Donnell (1992), Poklewski-Koziell (1992)
Falco tinnunculus	Kestrel	O'Donnell (1992), Poklewski-Koziell (1992)
Accipiter nisus	Sparrow hawk	O'Donnell (1992), Poklewski-Koziell (1992)
Buteo buteo	Buzzard	O'Donnell (1992), Poklewski-Koziell (1992)
Tyto alba	Barn owl	O'Donnell (1992), Poklewski-Koziell (1992)

favourite experimental material of Robinson, and Preston used, among many others, *Chaetomorpha* and *Cladophora*). But a notable feature of algal cell walls is that they show an impressive variety within the group (Neville, 1988c), including genera with helicoidal cell walls (Section 2.4b).

2.2b Annelid cuticle

A typical example of polychaete annelid cuticle with orthogonally oriented collagen fibres is *Alvinella pompejana*, a deep-sea (2,600 m) worm living in hydrothermal vents (Gaill et al., 1988). The fibres, however, also run a sinuous course, whereas in the cuticle of *Paralvinella grasslei* (Fig. 2.3), which is found in the same habitat, the collagen fibres are straight (Lepescheux, 1988).

2.2c Mantis eggcase protein

Female praying mantids surround each egg batch with a protein foam which eventually hardens to form an eggcase. Prior to secretion, the protein is in a liquid crystalline state; although it is mostly in a cholesteric (helicoidal) phase, some regions show an orthogonal texture (Fig. 2.4).

2.2d Rhinoceros beetle eggshell

The architecture of insect eggshells is extremely diverse (Hinton, 1981). That of *Oryctes rhinoceros* has an orthogonal layered structure of crystalline fibres in a matrix (Furneaux & Mackay, 1976). Truly orthogonal plies are very rare in arthropods.

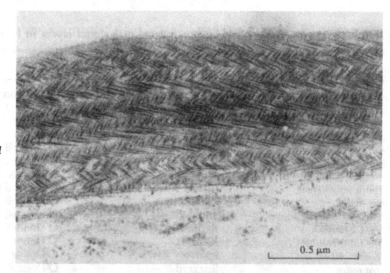

Figure 2.2. Electron micrograph of an oblique section through the cell wall of an alga, Eremosphaera viridis. This is a particularly clear example of a herring-bone pattern, which indicates an orthogonal fibre system: layers with parallel fibres set at right angles. From Weidinger and Ruppel (1985), by permission of the authors and Springer-Verlag, Wien.

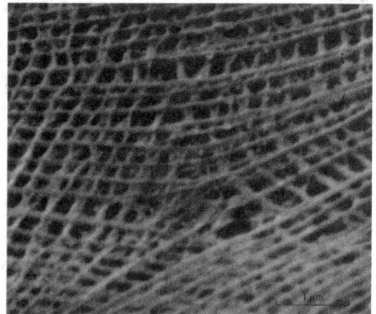

Figure 2.3. Scanning electron micrograph of predominantly orthogonal layers of collagen fibres seen in surface view of an annelid cuticle (Paralvinella grasslei). This worm lives in deep-sea hydrothermal vents. From Lepescheux (1988), by permission of the author and Editions Scientifiques Elsevier, Paris.

2.2e Sea cucumber defensive secretion

Holothuria forskali is an echinoderm which, if molested, produces a sticky defensive fluid. This contains collagen fibrils with 68-nm banding and is in a liquid crystalline state. Although mostly in the cholesteric (helicoidal) phase, some regions show an orthogonal texture (Fig. 2.5).

2.2f Vertebrate basement lamella

Epithelial cells cooperate to behave as a cohesive tissue because they sit on two layers which they have themselves secreted. These

Figure 2.4. Transmission electron micrograph of a section through the eggcase protein of a praying mantis (Sphodromantis tenuidentata). This was fixed in situ in the oothecal gland and was in a liquid crystalline phase prior to fixation. The system is seen to be orthogonal. From work by B. M. Luke and A. C. Neville. Stained with uranyl acetate and lead citrate.

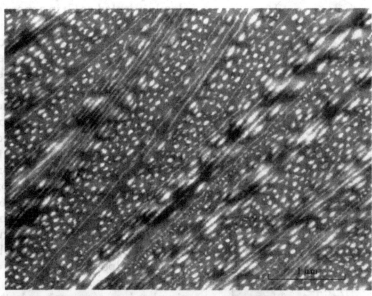

Figure 2.5. Transmission electron micrograph of a transverse section of a Cuvierian tubule from a sea cucumber (Holothuria forskali; Echinodermata, Holothuroidea). The collagen fibrils are seen alternately in longitudinal section (showing a 68-nm banding system) and in transverse section (showing characteristic irregular outline of fibrils). From Dlugosz et al. (1979), reprinted with permission from Maxwell Pergamon Macmillan plc, Oxford, and from the authors.

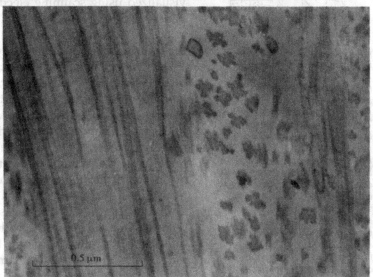

layers are a *basal lamina* (sometimes known as basement membrane) underlain by a *basement lamella* (also known as sub-basement membrane). Figure 2.6A illustrates these layers in diagrammatical form. In sponges, a primitive group, there is no basement lamella, and so their cells cannot function as coordinated tissues.

In the electron microscope the basal lamina appears diffusely granular, with filaments 4–6 nm in diameter. By contrast, the basement lamella consists of triple helical collagen fibrils set in a glycosaminoglycan matrix. The collagen lies more or less parallel

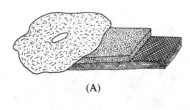

(A)

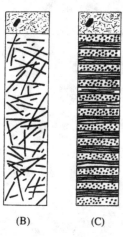

(B) (C)

Figure 2.6. Basement lamella – structure and regeneration. (A) Diagram of an epidermal cell sitting on laminar products secreted by itself and neighbours. The basement lamella is often a plywood of collagen fibrils. The basal lamina lies between the basement lamella and the basal plasma membrane of the cell; the basal lamina is granular or filamentous. The basement lamella can be revealed by enzymatic digestion of the overlying cells and basal lamina – a method used by Overton (1976). Although extracellular, the basement lamella plays a vital part in morphogenesis and wound repair. (B & C) Stages in the recovery after wounding of epidermal basement lamella in the skin of an amphibian axolotl (Amblystoma). At an early stage (B) the gap in the basement lamella is plugged with collagen fibrils, but these are randomly oriented. Later (C) they have become oriented into a twenty-three-ply laminated structure like that found in normal axolotls. Based upon electron micrographs by Weiss and Ferris (1956).

to the plane of the epithelium and may otherwise be (1) randomly oriented or (2) in a more or less orthogonal plywood (Figs. 2.7 and 2.8) or (3) helicoidal (as in flatworms, Fig. 2.20). The basal lamina is vitally important in cell adhesion, in positioning and anchorage of cells during development, and in recovery from wounding. It provides a prime example of the influence of a seemingly inert extracellular system upon nearby cells, both during normal development and during recovery from damage. The part played by self-assembly in the development of mammalian basal lamina is reviewed in detail by Yurchenco (1990).

The basement lamella beneath the epidermis of amphibians has proved popular material for studies on collagen fibril architecture. Collagen fibril directions were studied initially by an indirect method by Rosin (1946), who inferred the existence of an orthogonal arrangement from the positions of the pigmented processes of overlying melanophore cells. Overton (1976) exposed the underlying fibrils in the basement lamella of *Pseudacris triseriata* tadpoles by digestion of the overlying epidermal cells and basal lamina using trypsin. The fibril network could then be visualized using scanning electron microscopy. In a subsequent paper she first exposed the basal lamina of *Xenopus laevis* tadpoles with EDTA, and then removed it with hyaluronidase to reveal the collagen plies in the underlying basement lamella (Overton, 1979). The collagen fibril directions of tadpole epidermal basement lamella are precise. Typical layer directions for *Pseudacris* are shown in Figure 2.9 (Overton, 1976). The basic orthogonal pattern is modified according to the shape of the tadpole.

The paper by Weiss and Ferris (1956) on collagen fibril directions in the basement lamella of axolotl larvae (*Amblystoma*) during repair after wounding was for me a landmark. I became aware of the work through the influential book by Picken (1962) and read it in the early sixties when I was beginning work on chitin fibril directions in insect cuticles. It encouraged me to know that related work existed in the vertebrate field – I already knew of the extensive interest among botanists in cell wall cellulose microfibrils. In the basement lamella of normal axolotl skin there is a precise collagen plywood with twenty-three orthogonally oriented plies (Fig. 2.6C). After wounding, a gap in the basement lamella is plugged by a matrix with collagen fibrils, but these are initially arranged randomly (Fig. 2.6B). By a later stage of recovery the twenty-three-ply system becomes re-established. The astonishing thing is that the plies first become reorganized in those layers farthest away from the epidermal cells which secrete them; this mystery remains unsolved. Perhaps they align with the collagen at the edges of the wound?

The functions of basement membrane and lamella include mechanical support, semi-permeable filtration, creating a boundary between different types of cells, and influencing the position and

Figure 2.7. Electron micrograph of section through the tail epidermal basement lamella of a tadpole (Pseudacris triseriata; Amphibia, Anura). The superficial epidermal cells and basal lamina were removed with trypsin, to reveal the underlying epidermal basement lamella collagen fibres. The architecture is basically orthogonal, with changes in direction at an exceptional point (P). From Overton (1976), reprinted by permission of Wiley–Liss, a division of John Wiley and Sons, Inc., New York (copyright owner), and of the author.

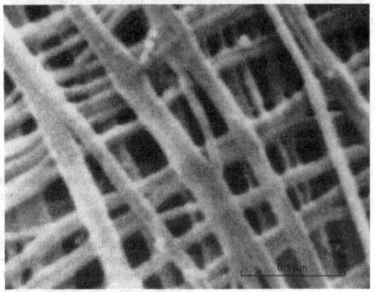

Figure 2.8. As for Figure 2.7, but from the cornea of the right eye of a Pseudacris triseriata tadpole. The cornea is equivalent to the basement lamella of the rest of the skin epidermis. The layers of collagen fibres are arranged as an orthogonal system which twists clockwise. From Overton (1976), reprinted by permission of Wiley–Liss, a division of John Wiley and Sons, Inc., New York (copyright owner), and of the author.

metabolism of embryonic cells which migrate over them (Fig. 3.28). There are important medical implications if they malfunction. They are involved in diabetes mellitus, glomerulonephritis, renal vein thrombosis, emphysema, and lipoid nephrosis, as well as third-degree burns. Whereas skin can recover from first- or second-degree burns (even if the Malpighian cell layer is damaged), it cannot recover from damage to the underlying and deeper-seated basement layers. Although these seem like innocuous extracellular structures, their damage leads to perma-

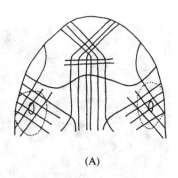

Figure 2.9. Diagrams of the orientations of collagen fibres in the epidermal basement lamella of the tadpole stage of a frog (Pseudacris). (A) dorsal view of head region. (B) Lateral view of tail region. Redrawn from Overton (1976).

nent scars on the skin. For a detailed review on basement membrane metabolism, see Kefalides, Alper, and Clark (1979). Several additional examples of orthogonal basement lamella are listed in Table 2.1A.

2.2g Fish scales

The scales of bony fish have recently been shown to be built of a twisted plywood of collagen fibres (Giraud et al., 1978). The rotation is very regular, especially in the coelacanth (*Latimeria chalumnae*). The system appears as if the building unit is a pair of orthogonal layers, with a right-handed rotation between it and the next pair. Oblique sections then generate two interleaved arced patterns (Fig. 2.10A), as if two helicoidal stacks of layers had been collated as odd and even layers – like a professional shuffle of playing cards (Fig. 2.10B). In coelacanths the sequence of angles between layers is 1–2 (90°), 2–3 (90° + 30°), 3–4 (90°), 4–5 (90° + 30°), and so on. In another primitive fish (*Amia calva*), however, the layer sequence (Fig. 2.11) twists in a left-handed direction (Meunier, 1981). The layers are exposed by fracture in liquid nitrogen, then are gold-coated and examined in a scanning electron microscope. In the roach (*Rutilus rutilus*), Darke (1986) found a left-handed twisted plywood (Fig. 2.12); in contrast, that of a lemon sole (*Pleuronectes microcephalus*) is right-handed (Fig. 2.13). Because there are no interconnecting helicoidal

2.2 Orthogonal plywoods and other large angular shifts / 35

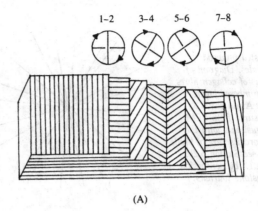

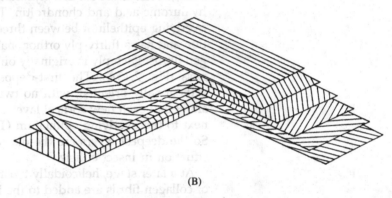

Figure 2.10. Orthogonal 'plywood' in fish scales. (A) Diagram of the twisted orthogonal layers of isopedin fibres in a coelacanth (Latimeria) scale. Layers 1 and 2 are nearest the reader and form an orthogonal pair. Subsequent pairs are rotated progressively clockwise, as shown by the arrows on the circles. The fibre directions are seen in oblique section, but the interleaving obscures the two sets of arced patterns. (B) As for part A, but with the even- and odd-numbered layers separated out to emphasize the two interleaved sets of arced patterns. Redrawn from Giraud et al. (1978).

layers, this twisted plywood structure in fish scales is different from the pseudo-orthogonal layers in the cuticle in some insects (beetles, bugs); compare, for example, Figure 1.17. The function of the twisted structure in fish scales is puzzling.

2.2h Cornea of vertebrate eyes

The corneas of fish (carp and goldfish), amphibia (frogs), a reptile (turtle), and some birds (duck, pheasant, quail) have a twisted orthogonal architecture of collagen fibrils (Trelstad, 1982) like that described for fish scales in the preceding section. An electron micrograph of a section through the cornea of a frog, *Rana temporaria*, is shown in Figure 2.14. Overton (1976) has shown that corneal structure may also be observed directly by scanning electron microscopy: That of another species of frog (*Pseudacris triseriata*) is shown in Figure 2.8. It has right-handed twisted orthogonal collagen layers.

As with fish scales and the corneas of fish, amphibia, and reptiles, the cornea in birds also has a complex building unit: an almost orthogonal pair of collagen layers. The *primary* stroma of the cornea in chicks (*Gallus domesticus*) contains no cells; the

36 / *The occurrence of fibrous composites*

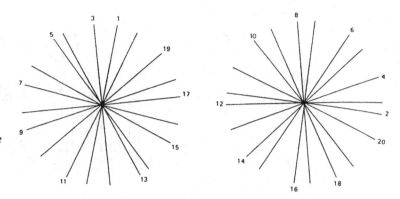

Figure 2.11. Twisted plywood structure in fish scales. Diagram of successive directions of collagen fibres in odd- and even-numbered layers of scales of the fish Amia calva. Directions are measured from SEM micrographs. Contrast this with the pseudo-orthogonal architecture of beetle cuticle (Fig. 5.22); see Section 5.2k. From Meunier (1981), by permission of Churchill Livingstone Journals, Edinburgh, U.K.

collagen fibrils are set in a glycosaminoglycan matrix, containing hyaluronic acid and chondroitin. The stroma is secreted by the overlying epithelium between three and ten days after fertilization and has a thirty-ply orthogonal scaffolding of striated collagen fibrils. Each ply is originally only one collagen fibril in thickness (Fig. 2.15A). The first-deposited (innermost) layers are orthogonally oriented, with no twist. Autoradiographic studies show that each new corneal layer of collagen fibrils is deposited next to the overlying epithelium (Trelstad & Coulombre, 1971). So the deepest layers are the oldest; this is the opposite of the situation in insect cuticle.

At a later stage, helicoidally twisted orthogonal pairs of layers of collagen fibrils are added to the inside face of the epithelium. Measured going away from the observer they each have a clockwise rotation of about 2° in left and right eyes alike. They are therefore (like arthropod cuticle and tunicate test) asymmetrical with respect to the whole body symmetry (Coulombre & Coulombre, 1975). These helicoidally twisted layers therefore end up nearest to the outer surface of the eye. The patterns generated in sections by the orthogonal-to-helicoidal transition are discussed in Section 5.1d. Figure 2.16 shows plots of the angular rotation of corneal collagen in the eyes of a number of vertebrates (Trelstad, 1982).

The primary stroma is invaded by wandering fibroblast cells, which secrete more collagen. The additional layers of collagen fibrils take up the same directions as the original fibrils of the primary stroma. So the inner orthogonal system and the outer helicoidal system finally consist of component plies, each of which contains numerous collagen fibrils (Fig. 2.15B).

I echo the plea by Hay (1980) for extended comparative studies on structure and morphogenesis in a selection of bird species. The results might relate to visual accuracy and acuity. It has been suggested that the helicoidal optics could even function as an analyzer for detection of plane-polarized light from the sky

Figure 2.12. Scanning electron micrograph of a twisted orthogonal plywood system seen in fish scales. This is from a roach (Rutilus rutilus; Cyprinidae). The layers were exposed by fracture in liquid nitrogen. They have a left-handed sense of rotation. Courtesy of Wendy Darke (1986).

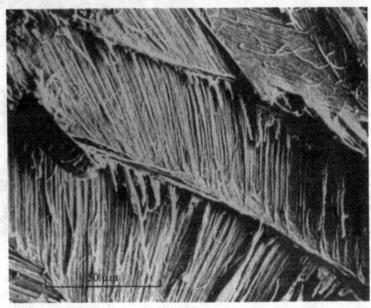

Figure 2.13. As for Figure 2.12, but for the scales of a lemon sole (Pleuronectes microcephalus; Pleuronectidae). By contrast with the previous species, the layers rotate clockwise. Courtesy of Wendy Darke (1986).

(Hay, 1980); this is considered in Section 3.3e. Further research on cornea will be crucial to an understanding of various forms of vision impairment in humans. The orthogonal pattern in mammals is recorded as different from that in birds (Ozanics, Rayborn, & Sagun, 1977). The mammalian primary stroma is not so well organized and contains only a few orthogonally arranged fibrils (Hay, 1980).

38 / *The occurrence of fibrous composites*

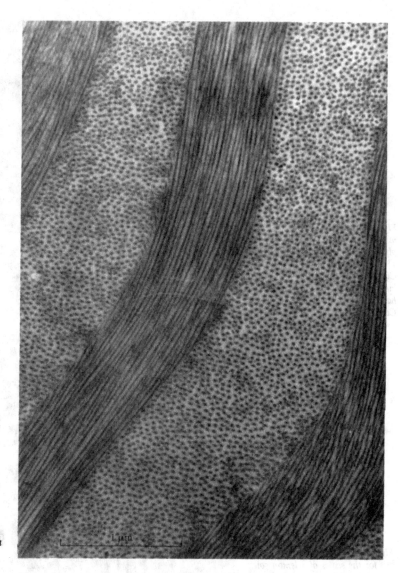

Figure 2.14. Electron micrograph of a section through the cornea of an amphibian eye (common frog, Rana temporaria). The collagen fibrils form a neat orthogonal plywood. Micrograph courtesy of the late Dr. J. N. Lythgoe.

Recent work by two of my students has greatly extended the list (Table 2.1B) of birds known to have twisted plywood layers of collagen in their corneas (O'Donnell, 1992; Poklewski-Koziell, 1992). The abundance of this type of architecture in vertebrate corneas implies that it may be serving some important function.

2.2i Human bone

The classical interpretation of human bone is of stacks of orthogonal layers of collagen fibrils, rolled into co-axial cylinders (osteons), which are then grouped in parallel (Gebhardt, 1906).

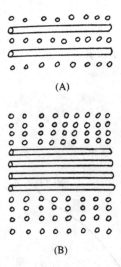

Figure 2.15. *Plywood thickening in bird cornea.* (A) In the cornea of chicken eye, an orthogonal ply of collagen fibrils is initially established. Each ply is originally only one collagen fibril in thickness. (B) At a later stage, fibroblast cells invade and secrete additional layers which follow these same directions, as shown by Bard and Higginson (1977). Each original ply thus comes to contain several layers of collagen fibrils.

2.3 Helicoidal plywoods in animals

Transmission electron micrographs of this orthogonal type of collagen scaffolding in human bone are given by Giraud-Guille (1988). She confirmed the orthogonal interpretation by absence of change of image in tilting experiments (see Section 1.3k). Giraud-Guille (1988) has also convincingly demonstrated helicoidal osteons in human bone (Section 2.3q).

2.2j Large angles other than 90°

Parallel fibres are oriented mutually at 0°; orthogonal layers are oriented at 90°. All other possible angles may exist between these limits. They exist in a wide selection of systems and may be regarded as special cases of helicoids. A typical example is found in the cuticle of the recently discovered phylum of deep-sea worms which live in hydrothermal vents (Vestimentifera). In *Riftia pachyptila*, successive layers of fibres in the cuticle are oriented through angles ranging from 30° to 60° (Gaill et al., 1988).

The egg capsule of the lesser-spotted dogfish (*Scyliorhinus caniculus*) is known as the mermaid's purse. It is usually attached to seaweed by means of four twisted tendrils. Most examples found washed up on the shore have already hatched and are empty. The inner layers of the egg capsule wall consist of fibres of collagen-like protein arranged in parallel sheets. The sheets, which are about 0.4 µm thick, are set at 45° (Knight & Hunt, 1974b).

Graptolites are an extinct group of marine colonial organisms, widely distributed in the seas during the early Paleozoic. They form one class (Graptolithina) of the phylum Hemichordata, itself closely related to the Chordata. Well-preserved fossil graptolites may be chemically isolated and then sectioned for transmission electron microscopy. Surprisingly, the outer cortex of *Dictyonema* contains several well-preserved plies of fibrils arranged at about 45°. Even more surprisingly, each fibrillar ply is partitioned from its neighbour by a membranous structure (Towe & Urbanek, 1972).

In humans, the layers of collagen fibres in the walls of intervertebral discs are oriented at 65° to the vertebral axis. The functional significance of this is described in Section 3.2j. If considered as a continuous rotation, this gives the curious result that the layers change direction alternately through 130° or 50°. An interesting developmental problem therefore arises.

Helicoids generate ranks of arced patterns (Fig. 1.8). Historically, such patterns were first drawn by Schulze (1863) from observations of cellulose (tunicin) fibres in sections of the test of a sea squirt (*Halocynthia papillosa*). They were, however, first fully

40 / *The occurrence of fibrous composites*

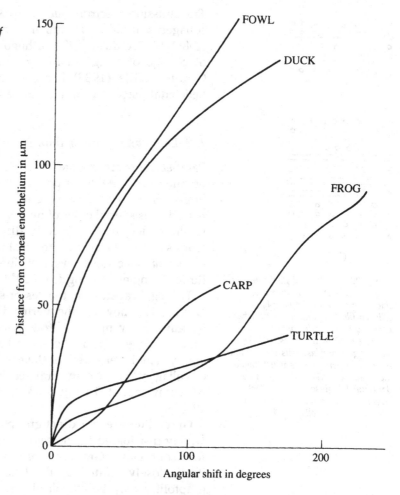

Figure 2.16. Plots of helicoidal rotation of collagen fibres in the eyes of various non-mammalian vertebrates. The innermost layers of the cornea are the oldest. These deepest layers have collagen fibrils arranged in an orthogonal ply. They are alternately parallel and perpendicular to the dorso-ventral axis of the eye. In layers deposited later (closer to the outer surface of the eye) the ply rotates with a progressive shift in angle. This sense of shift is clockwise in layers farther and farther from the observer. (It does not matter whether the observer is looking through the cornea from in front or from behind – the sense of shift of fibril orientation will still be clockwise.) The early and deeper layers do not rotate; the later ones do. (This point is sometimes confused in the literature, where the rate of shift is described as slowing in the deeper layers.) Data from only one of each successive pair of orthogonal layers are plotted on the graph; the other points would be displaced by 90°. The angle of each alternate collagen layer is measured with reference to the dorso-ventral axis of the eye. The collagen shows an angular (helicoidal) shift in the outermost layers in all the kinds of vertebrates shown on the graph. All are clockwise, and moreover they are clockwise in both left and right eyes. Replotted from Trelstad (1982), with curve for fowl (Gallus domesticus) added from Coulombre and Coulombre (1975). The species of duck, carp, frog, and turtle are not named in the original.

understood by W. J. Schmidt (1924), who had an enviable command of the use of polarizing optics. [Polarized light is cheap and easy to use, but there are pitfalls in its interpretation; see Neville (1980).] Schmidt reasoned that this test could not consist solely of orthogonal layers of fibrils set at right angles; otherwise it would be possible to find a plane of section at 45° to both sets of fibres such that these would not be distinguishable between crossed polaroids. What he actually observed were bright and dark layers in alternation in all sections cut vertically to the sheets of fibres. Schmidt reasoned that there must be a *gradual* change of fibre orientation from layer to layer.

By far the most influential landmark in helicoidal studies was the paper on crab cuticle by Yves Bouligand (1965). This has generated a large number of publications and widespread interest. My own interest in helicoidal systems arose from the receipt of a copy of Bouligand's paper. I was also influenced by a visit to Cambridge, where Lawrerce Picken drew my attention to the 'feathery' drawing of *Halocynthia* by Schulze (1863). I owe a great debt of gratitude to both Bouligand and Picken. The helicoidal principle helped me to interpret in ultra-structural terms my own experiments on locust cuticle.

It has emerged that helicoids are the most abundant regular architecture in extracellular fibrous composite materials. This is documented by the examples cited in Tables 2.2–2.6. The animal examples are divided into sections: insect cuticles, cuticles of other arthropods, moth eggshells, other invertebrates, and vertebrates. General reviews on helicoids in animals have been published by Bouligand (1972) and Neville (1975a).

2.3a Protistan cysts

Some freshwater protistans encyst at certain stages of their lives. These cyst walls, which may be thick or thin, can be very complicated, with as many as eight different kinds of layers deposited in sequence. Cysts of the heliozoan *Actinophrys sol* are brown and spherical, 20–30 µm in diameter and coated with silicated scales (Fig. 2.17). Some cyst walls contain a helicoidal layer, detected by arced patterns in electron micrographs [e.g. *Actinophrys sol* (Patterson, 1979) and *Echinosphaerium nucleofilum* (Patterson & Thompson, 1981)].

The usual functions of cysts are reproductive, digestive, or protective. They may serve in dispersal between temporary freshwater habitats or between hosts in parasitic forms and may be either short-lived (twenty-four hours) or long-lived (years), or else may act as a dormant stage, with reduced metabolic activity. Cysts of *Actinophrys sol* are unable to withstand desiccation. However, they need not dry out to serve their function. They may encounter other problems, such as periods of prolonged low temperature or poor nutrient supply; sinking to the bottom of a pond and encysting provides a solution (Dr. R. M. Crawford, personal communication).

Newman (1990) has identified specific types of intracellular vesicles which correlate with the eight kinds of cyst wall layers. In the case of the helicoid (layer 6) the vesicle contents are fibrillar, but they do not appear helicoidal until released outside the cell plasma membrane by exocytosis. No microtubules were detected during cyst wall formation as the axopodia depolymerized prior to layer deposition. Microtubules are therefore not necessary for helicoidal morphogenesis in *Actinophrys* cyst wall.

2.3b Sponge gemmules

Electron microscopy of oblique sections through sponge gemmule walls reveals arced patterns indicative of helicoidally organized

Table 2.2. *A selection of insects for which electron micrographs of cuticle show arced patterns which indicate microfibrils oriented in helicoids. The purpose is to show how helicoidal cuticle is widespread throughout the insect orders. Unless otherwise indicated, the examples in this table refer to adult cuticles.*

Order	Species	Reference
Thysanura	*Monachina schultzei*	Caveney (1969)
Collembola	*Podura aquatica*	Noble-Nesbitt (1963)
Ephemeroptera	*Coluburiscoides* (nymph)	Filshie & Campbell (1984)
Odonata	*Aeschna juncea*	Neville (1975a)
Dictyoptera	*Periplaneta americana*	Neville & Luke (unpublished)
	Blaberus craniifer	Neville & Luke (unpublished)
Phasmida	*Carausius morosus*	Neville & Luke (1969b)
Mantodea	*Sphodromantis tenuidentata*	Neville & Luke (unpublished)
	Sphodromantis centralis (eggcase proteins)	Kenchington & Flower (1969)
	Sphodromantis tenuidentata (eggcase proteins)	Neville & Luke (1971b)
	Miomantis monarcha (eggcase proteins)	Neville & Luke (1971b)
Orthoptera	*Acheta domesticus*	Neville & Luke (unpublished), Hendricks & Hadley (1983)
	Schistocerca gregaria	Neville & Luke (1969a)
	Locusta migratoria	Neville & Luke (unpublished)
	Romalea microptera	Brown (1987), Banerjee (1988a)
Hemiptera	*Belostoma malkini*	Neville & Luke (1969b)
	Hydrocyrius colombiae	Neville & Luke (1969b)
	Rhodnius prolixus	Lai-Fook (1968)
Isoptera	*Cubitermes fungifaber* (female)	Bordereau (1982)
	Macrotermes bellicosus (female)	Bordereau (1982)
Diptera	*Drosophila melanogaster* (pupa)	Wolfgang et al. (1986)
	Lucilia cuprina (larva)	Filshie (1970)
	Chironomus riparius (larva)	Credland (1983)
	Glossina morsitans	Neville & Luke (unpublished)
	Sarcophaga bullata (larva)	Neville & Luke (unpublished)
	Aedes aegypti	Chevone & Richards (1977)
Lepidoptera	*Amauris ochlea* (pupa)	Steinbrecht (1985)
	Aglais urticae	Neville & Luke (unpublished)
	Danaus plexipus	Neville & Luke (unpublished)
	Calpodes ethlius (larva, pupa)	Locke (1960)
	Galleria mellonella (larva)	Locke (1961)
	Macroglossum stellarum	Neville & Luke (unpublished)
	Antheraea pernyi (larva)	Neville & Luke (unpublished)
	Hyalophora cecropia (pupa)	Neville & Luke (unpublished)
	Zygaena trifolii	Neville & Luke (unpublished)
	Manduca sexta (larva)	Wolfgang & Riddiford (1986)
Hymenoptera	*Apis mellifera*	Locke (1961)
Coleoptera	*Tenebrio molitor* (larva, pupa)	Neville & Luke (1969b), Delachambre (1975)
	Lampyris noctiluca	Caveney, in Neville (1975a)
	Plusiotis resplendens	Caveney (1971)
	Xyleborus ferrugineus	Chu et al. (1975)
Strepsiptera	*Elenchus tenuicornis*	Kathirithamby et al. (1990)
Siphonaptera	*Xenopsylla cheopsis*	Neville & Luke (unpublished)

Table 2.3. *Selection of arthropods (excluding insects) for which electron micrographs of cuticle show arced patterns which indicate microfibrils oriented in helicoids*

Group	Species	Reference
Phylum Crustacea		
Ostracoda	*Cypridopsis vidua*	Bate & East (1972)
Branchiopoda	*Triops cancriformis*	Rieder (1972)
	Artemia salina (cyst wall)	Morris & Afzelius (1967)
Phyllopoda	*Tanymastix* sp.	Bouligand (1972)
Amphipoda	*Gammarus pulex*	Halcrow (1978)
	Hyale nilssoni	Halcrow (1985)
Copepoda	*Cletocampus retrogressus*	Gharagozlou & Bouligand (1973)
	Acanthocyclops viridis	Bouligand (1965)
	Copepod spermatophore	Gupta (personal communication)
Malacostraca	*Panulirus argus*	Neville & Luke (1971a), Filshie & Smith (1980)
	Astacus fluviatilis	Neville (1975a)
	Carcinus maenas	Bouligand (1965), Neville (1970)
	Homarus vulgaris	Neville & Luke (unpublished)
	Homarus americanus	Arsenault et al. (1984)
	Oniscus asellus	Neville & Luke (unpublished)
	Eryma stricklandi (fossil)	Neville & Berg (1971)
	Cancer pagurus	Bouligand (1965)
	Orconectes viridis	Travis & Friberg (1963)
	Cambarus bartonii	Halcrow (1988)
	Crangon septemspinosa	Halcrow (1988)
	Gammarus oceanicus	Halcrow (1988)
	Idotea baltica	Powell & Halcrow (1985)
Phylum Chelicerata		
Scorpionida	*Hadrurus arizonensis*	Filshie & Hadley (1979)
Xiphosura	*Limulus polyphemus*	Neville (1967c)
Araneida	*Tegenaria agrestis*	Neville & Luke (unpublished)
	Aphonopelma chalcodes	Cooke, in Neville (1975a)
	Cupiennius salei	Barth (1969, 1970)
	Latrodectus hesperus	Hadley (1981)
Phylum Antennata		
Diplopoda	*Rhinocricus nodulipes*	Silvestri (1903)
	Julus sp.	Neville (1975a)
	Orthoporus ornatus	Walker & Crawford (1980)
Chilopoda	*Lithobius forficatus*	Neville (1975a)

layers. These are known from *Ephydatia fluviatilis* (De Vos, 1972). There is scope for a wider survey of helicoidal structure in sponges.

2.3c Flatworm basement lamella

The epidermal cells of flatworms (Platyhelminthes: class Turbellaria) sit on an extracellular basement lamella. In some species this is known to have helicoidal structure (Table 2.5), as indicated by ranks of arced patterns (Fig. 2.20). So far, these have been

Table 2.4. *Butterfly and moth eggshells for which electron micrographs show arced patterns indicating helicoidal proteins in the chorion*

Species	Family	Reference
Hyalophora cecropia	Saturniidae	Telfer & Smith (1970), Smith et al. (1971)
Antheraea polyphemus	Saturniidae	Regier et al. (1982)
Bombyx mori	Bombycidae	Furneaux & Mackay (1972), Matsuzaki (1968)
Pieris brassicae	Pieridae	Furneaux & Mackay (1972), Matsuzaki (1968)
Laspeyresia pomonella	Olethreutidae	Furneaux & Mackay (1972), Matsuzaki (1968)
Lymantria dispar	Lymantriidae	Furneaux & Mackay (1972), Matsuzaki (1968)
Chilo suppressalis	Crambidae	Furneaux & Mackay (1972), Matsuzaki (1968)
Calpodes ethlius	Hesperiidae	Griffith & Lai-Fook (1986)
Phalera bucephala	Ceruridae	Chauvin et al. (1974)
Acrolepia assectella	Plutellidae	Chauvin et al. (1974)
Plutella maculipennis	Plutellidae	Chauvin et al. (1974)
Manduca sexta	Sphingidae	Regier & Vlahos (1988)

Table 2.5. *Invertebrates (excluding arthropodan phyla) for which electron micrographs of various skeletal materials show arced patterns which indicate microfibrils oriented as helicoids*

Phylum	Species	Structure	Reference
Protista	Echinosphaerium nucleofilum	Cyst wall	Patterson & Thompson (1981)
	Actinophrys sol	Cyst wall	Patterson (1979)
	Arcella	Shield	Weis-Fogh (personal communication)
	Radiolarian sp.	Capsule	Lecher & Cachon (1970)
Porifera	Ephydatia fluviatilis	Gemmule wall	De Vos (1972)
Cnidaria	Aurelia aurita	Podocyst cuticle	Chapman (1968)
Platyhelminthes	Discocelides langi	Basement lamella	Pedersen (1966)
	Stylochus sp.	Basement lamella	Tyler (1984)
	Kytorhynchella meixneri	Basement lamella	Rieger (1981)
Annelida	Diopatra cuprea	Oocyte envelope	Anderson & Huebner (1968)
	Alvinella pompejana	Dwelling tube	Gaill & Hunt (1986)
	Erpobdella octoculata	Egg capsule	Knight & Hunt (1974a)
Vestimentifera	Riftia pachyptila	Cuticle	Gaill et al. (1988)
Rotifera	Philodina roseola	Integument	Dickson & Mercer (1967)
	Asplanchna brightwelli	Integument	Brodie (1970)
	Notommata sp.	Integument	Clément (1969)
	Abrochta sp.	Integument	Clément (1977)
Nematoda	Mermis nigrescens	Cuticle	Lee (1970)
	Globodera rostochiensis	Cuticle	Shepherd et al. (1972)
	Trichuris suis	Eggshell chitin	Wharton (1978)
	Capillaria hepatica	Eggshell chitin	Grigonis & Soloman (1976)
	Atalodera ucri	Female cuticle	Baldwin (1983)
	Atalodera lonicerae	Female cuticle	Baldwin (1983)
	Sarisodera hydrophila	Female cuticle	Baldwin (1983)
	Cactodera cacti	Female cuticle	Baldwin (1986)
	Punctodera punctata	Female cuticle	Baldwin (1986)
	Thecavermiculatus gracililancea	Female cuticle	Baldwin (1986)

Table 2.5. *(cont.)*

Phylum	Species	Structure	Reference
Pogonophora	*Nereilinum punctatum*	Dwelling tube	Gupta & Little (1975)
Mollusca	*Buccinum undatum*	Periostracum	Hunt & Oates (1970, 1978)
		Egg capsule	Hunt (1971)
	Pterocyclus latilabrum	Periostracum	Hunt & Oates (1985)
Echinodermata	*Thyone* sp.	Body wall	Gross & Piez (1960)
	Holothuria forskali	Cuvierian tubule fluid	Dlugosz et al. (1979)
	Havelockia inermis	Connective tissue	Bouligand (1972, 1978b)
Chordata (Tunicata)	*Halocynthia papillosa*	Test tunicin	Schulze (1863), Neville (1967c), Bouligand (1972), Gubb (1975)

Note: For vertebrate examples, see Table 2.6.

Table 2.6. *Species of vertebrates for which electron micrographs indicate helicoidal structure, either (A) as arced patterns or (L) as laminations*

Species	Reference
Fish	
(A) *Cynolebias belotti* (eggshell)	Sterba & Müller (1962)
(A) *Agonus cataphractus* (eggshell)	Götting (1965, 1967)
(A) *Merlangius merlangus* (eggshell)	Götting (1967)
(A) *Lutjanus analis* (eggshell)	Erhardt (1978)
(A) *Salmo salar* (eggshell)	Hunt & Oates (1978)
(A) *Gadus morrhua* (eggshell)	Grierson & Neville (1981)
(A) *Pleuronectes platessa* (eggshell)	Grierson & Neville (1981)
(A) *Salmo gairdneri* (eggshell)	Grierson & Neville (1981), Hamodrakas et al. (1987)
(A) *Limanda limanda* (eggshell)	Lönning (1972)
(L) *Crenilabrus melops* (eggshell)	Lönning (1972)
(L) *Crenilabrus cinereus* (eggshell)	Lönning (1972)
(L) *Crenilabrus mediterraneus* (eggshell)	Lönning (1972)
(L) *Crenilabrus exoletus* (eggshell)	Lönning (1972)
(L) *Crenilabrus rupestris* (eggshell)	Lönning (1972)
(L) *Platichthys flesus* (eggshell)	Lönning (1972)
(L) *Hippoglossoides platessoides* (eggshell)	Lönning (1972)
(A) Goldfish, sp. not cited (corneal collagen)	Trelstad (1982)
(A) Carp, sp. not cited (corneal collagen)	Trelstad (1982)
Amphibia	
(A) *Pseudacris triseriata* (corneal collagen)	Overton (1976)
(A) Frog, sp. not cited (corneal collagen)	Trelstad (1982)
Reptilia	
(A) Turtle, sp. not cited (corneal collagen)	Trelstad (1982)
Birds	
(A) *Gallus domesticus* (chick) (corneal collagen)	Coulombre & Coulombre (1961, 1975)
(A) Pheasant, sp. not cited (corneal collagen)	Trelstad (1982)
(A) Quail, sp. not cited (corneal collagen)	Trelstad (1982)
(A) Duck, sp. not cited (corneal collagen)	Trelstad (1982)
Mammals	
(A) *Homo sapiens* (leg bone collagen)	Chambers et al. (1984), Giraud-Guille (1988)

46 / *The occurrence of fibrous composites*

Figure 2.17. Scanning electron micrograph of a cyst from a freshwater protist (Actinophrys sol). One of the many layers in the cyst wall is helicoidal (Patterson, 1979). Reproduced by kind permission of Dr. J. Newman (1990).

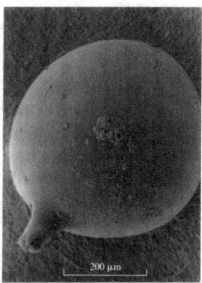

Figure 2.18. Scanning electron micrograph of a female cyst of a nematode worm: the potato root eelworm (Globodera rostochiensis). Only females encyst, changing shape from cylindrical to spherical, and adding an extra (helicoidal) layer of collagen fibrils to the inside of the cyst wall cuticle (Shepherd et al., 1972). These cysts are attached initially by the sucking mouthparts (projection on photograph) to the underground stems of potatoes. Photographed in Bristol University Zoology Department from material supplied by Dr. B. Kerry of Rothamsted Experimental Station.

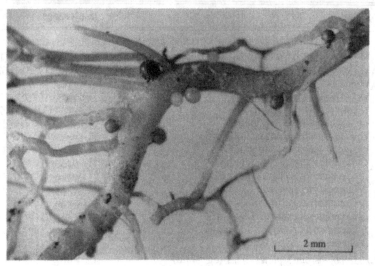

Figure 2.19. Photograph of potato root eelworm cysts (Globodera rostochiensis). The female cysts are attached initially by the sucking mouthparts to the underground stems of potatoes (on which they are a serious pest). They may subsequently drop off and lie dormant for twenty, years in the soil. These cysts are on a culture of potato cultivar Désiree. Photograph courtesy of Dr. C. C. Doncaster (Rothamsted Experimental Station, Hertfordshire, U.K.).

seen accidentally in oblique sections; the original authors have been unaware of their significance. The time is right for a deliberate investigation. (Basement lamella is an important field of research in vertebrates; see Section 2.2f.) The clear electron micrographs of *Kytorhynchella* (Rieger, 1981) in Figure 2.20 show 1.5 turns of helicoid.

2.3d Nematode cysts

Nematode worms are a major group of plant pathogens of economic importance. Helicoidal structure has recently been found

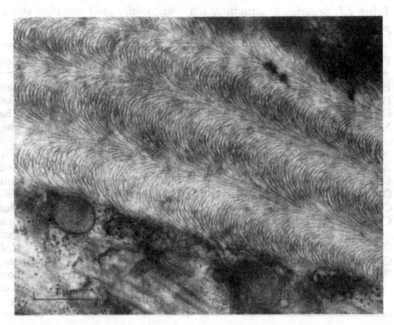

Figure 2.20. Transmission electron micrograph of a section through the helicoidal basement lamella of the epidermis in a flatworm (Kytorhynchella meixneri); Platyhelminthes, Turbellaria). The basement lamella shows arced patterns indicative of 540° rotation of helicoidal layers. Photograph from Rieger (1981); reprinted by permission of Kluwer Academic Publishers and the author.

in the cyst walls of some genera of Heteroderidae and is even being used to test hypotheses on their phylogeny (Cliff & Baldwin, 1985).

Characteristic ranks of arced patterns were first seen in electron micrographs of sections through the cysts of potato root eelworms, *Globodera* (formerly *Heterodera*) *rostochiensis*, by Shepherd, Clark, and Dart (1972). They approached me for help in interpretation. Only females change shape by rounding off from their original cylindrical form to make cysts (Fig. 2.18), and only females contain helicoidal layers. These are collagenous and are added to form the innermost layer of the cyst wall. The helicoidal layer in *G. rostochiensis* is 6–8 μm thick. The cysts of this species are more or less spherical and measure 0.5–0.8 mm in diameter. They are embedded by the mouthparts into the outer surfaces of potato roots (Fig. 2.19). Females enlarge to become up to 100 times larger than males. The ovaries enlarge to produce the enormous number of eggs characteristic of an endoparasitic life-style. When mature, cysts detach from the roots. The cyst serves to enclose and protect the eggs and may lie dormant in the soil for twenty years. This makes control difficult, as the cysts may be spread in soil from infested areas. That the helicoidal layer is found only in females, and not in the body wall of male worms, would indicate that it is serving a protective function; using helicoidal structure is a strong way to build a spherical shell. The same principle is used in many other types of spores and eggshells (Table 2.7).

Table 2.7. *Selection of structures with helicoidally built spherical shells, as indicated by parabolic patterning in electron micrographs of oblique sections*

Phylum	Species	Structure	Reference
Protista	*Actinophrys sol*	Cyst wall	Patterson (1979)
Porifera	*Ephydatia fluviatilis*	Sponge gemmule wall	Vos (1974)
Cnidaria	*Aurelia aurita*	Jellyfish podocyst	Chapman (1968)
Nematoda	*Globodera rostochiensis*	Potato root eelworm (female cyst wall)	Shepherd et al. (1972)
	Trichuris suis	Pig eelworm eggshell	Wharton (1978)
Annelida	*Diopatra cuprea*	Bristleworm eggshell	Anderson & Huebner (1968)
Lepidoptera	*Hyalophora cecropia*	Silk moth eggshell	Smith et al. (1971)
Orthoptera	*Melanoplus differentialis*	Grasshopper serosal cuticle	Slifer & Sekhon (1963)
Crustacea	*Artemia salina*	Brine shrimp eggshell	Morris & Afzelius (1967)
Chordata	*Gadus morrhua*	Codfish eggshell	Grierson & Neville (1981)
	Cynolebias belotti	Killifish eggshell	Sterba & Müller (1962)
Chlorophyta	*Cylindrocapsa geminella*	Green alga zoospore	Hoffman & Hofmann (1975)
Charophyta	*Chara vulgaris*	Stonewort oospore	Baynes (1972), Neville et al. (1976)
Fungi	*Glomus epigaeum*	Spore wall	Bonfante-Fasolo (1983), Bonfante-Fasolo & Vian (1984), Bonfante-Fasolo & Grippiolo (1984)

Females of some genera do not form cysts and yet produce the helicoidal layer (known to taxonomists as the D-layer); these include *Atalodera ucri*, *A. lonicerae*, and *Sarisodera hydrophila* (Baldwin, 1986). The following also have the helicoidal layer (Cliff & Baldwin, 1985): *Globodera rostochiensis* and *G. pallida* (on potato), *G. tabacum* (on tobacco), *Cactodera* (on ornamental cacti), *Punctodera punctata* (on wheat), and *Thecavermiculatus gracililancea*. The helicoidal layer is absent from *Heterodera* (*sensu stricto*), whose lemon-shaped cysts may be distinguished from the round cysts of *Globodera*. Hence there are no helicoids in *H. glycines* (on soybean), *H. schachtii* (on sugar beet), *H. avenae* (on oats), or *H. betulae*. Helicoids are also absent from *Sarisodera africana*, which has therefore been transferred to *Afenestrata africana*.

In a whole volume devoted to cyst nematodes (Lamberti & Taylor, 1986) there is surprisingly no mention either of arced patterns or of their helicoidal interpretation. I would suggest that more widespread recognition of the helicoidal feature could shed further light on nematode phylogeny (Baldwin, 1986) and on possible structural adaptation to semi-arid conditions (Shepherd et al., 1972; Baldwin, 1983).

2.3e Nematode eggshells

The only place in the nematode life cycle where chitin is found is in the eggshell. In *Trichuris suis*, a pig parasite, a helicoidal layer of chitin microfibrils set in a protein matrix is sandwiched between an outer vitelline layer and an inner lipid layer (Wharton & Jenkins, 1978). Such eggs may remain viable on the ground for up to six years, and a helicoidal construction seems a strong way to build such persistent eggshells. *Trichuris* eggs are barrel-shaped and measure roughly 80 µm by 40 µm. Wharton and Jenkins (1978) and Wharton (1978) give convincing evidence that the eggshell contains a helicoidal fibrous composite of chitin and protein. The arced patterns seen previously in eggshells of *Capillaria hepatica* (Grigonis & Soloman, 1976) may be reinterpreted as arising from helicoidal layers. An opercular plug at either end of the egg is also helicoidal, but with wider pitch. This could facilitate the eventual hatching of the egg. The narrow arced pattern of the eggshell is in continuity with the broader arcs of the plug regions.

The collagenous helicoidal layers of nematode cysts occur in some genera and not in others (Section 2.3d), and the same appears to hold for nematode eggshells, although information is needed on a wider selection of material. Thus, helicoidal layering is found in *Trichuris* and *Capillaria*, but not in *Ascaris lumbricoides*, *Heterakis gallinarum*, or *Porrocaecum ensidicaudatum* (Wharton, 1980). Neither is it found in *Globodera rostochiensis*, although randomly oriented chitin microfibrils are present (Perry, Wharton, & Clarke, 1982). Thus, based on this small sample, helicoidal structure is present in the order Trichurida, but absent from the orders Oxyurida, Ascaridida, and Tylenchida (Wharton, 1980).

2.3f Rotifer cuticle

I interpreted the ranks of arced patterns seen in electron micrographs (Dickson & Mercer, 1967) of the integumental shield of a rotifer (*Philodina roseola*) as arising from helicoidal structure (Neville, 1975a). Several more uninterpreted examples in other genera of rotifers have since been published (Table 2.5).

2.3g Annelid eggshell

There are few observations indicating helicoidal structure in annelid worms. Sections of the eggshell of a polychaete (*Diopatra cuprea*) show arced patterning in electron micrographs (Anderson & Huebner, 1968); these have been interpreted as helicoidal (Bouligand, 1972). Knight and Hunt (1974a) found helicoidal structure in the egg capsule of a leech (*Erpobdella octoculata*); this surrounds the fertilized eggs.

2.3h Arthropod cuticles

It is probable that all arthropod cuticles contain helicoidal layers of chitin set in a matrix of proteins. In this one group alone helicoids thus occur in huge numbers of species and individuals (Table 2.2 for insects; Table 2.3 for other arthropods).

In addition to helicoids revealed by arced patterns in electron micrographs, they are also detected by laminations resolved in the light microscope since a correlation between the two has been established (Fig. 1.20). It is also known that helicoids of appropriate pitch reflect circularly polarized light in some scarab beetles (Neville & Caveney, 1969). Circularly polarized light may quickly be detected by means of simple optical devices, and this has enabled us to establish the presence of helicoidal structure in several thousand species of scarab beetles in the British Museum of Natural History. This was a real piece of luck; as entomologists we sometimes tend to compensate for the large number of insect species by overgeneralizations. At least in this case we had plenty of data!

The external cuticle dominates the ways of life of insects (Neville, 1978). It consists of a laminated exoskeleton (Fig. 5.20) secreted on the outer face of a single layer of epithelial cells (epidermis) which sits on an underlying basement lamella. The outermost layer (epicuticle) is thin (1–2 μm) and in terrestrial insects is waterproofed by lipids. Within this, the bulk of the cuticle is divided into exocuticle (secreted before a moult) and endocuticle (secreted to the inside of the exocuticle following a moult). The exocuticle contains chitin microfibrils (crystallites) of very small diameter (2.8 nm), which is nearly three times thinner than a cell membrane. These form a helicoid of small pitch, not usually resolvable in a light microscope. In the endocuticle, the pitch is often larger, so that laminations due to the helicoidal architecture can be seen with a light microscope. Each 180° rotation of helicoid creates one lamella (Fig. 1.20). The helicoidal components in some Crustacea (crayfish, crabs) are not the individual chitin crystallites, but bundles of them with a diameter of 25–50 nm (Fig. 2.21). In very oblique sections of such material the helicoidal pitch is sometimes so large that arced patterns can be seen using only a light microscope. Electron micrographs of arced patterns are shown for strepsipteran endocuticle (Fig. 2.22) and crab endocuticle (Figs. 2.21 and 2.23).

In insects, the proteins which form the matrix of the fibrous composite cuticle are stiffened by small, reactive aromatic molecules which cross-link protein chains together. The exocuticle is the most hardened layer and is often darkened as well. Crustacean cuticle resembles that of insects in general architecture. It contains, however, far less protein and is in many cases stiffened by crystalline calcium carbonate. Crustaceans secrete, in sequence, the epicuticle, exocuticle, and endocuticle, all of which are calcified extracellularly. They are underlain by a non-calcified (and badly named) membranous layer. The special interest in crustacean cuticle lies in its differences and similarities with calcified

Figure 2.21. Electron micrograph of an oblique section of endocuticle of a shore crab (Carcinus maenas; Crustacea, Malacostraca). The arced pattern is characteristic of helicoidal architecture. There are two coincident patterns here: (1) The arcs arise from the fibrillar component, and (2) the crescents derive from the pore canals (cellular extensions) which run through the thickness of the cuticle. Reproduced by kind permission from A. Plumptre (1987).

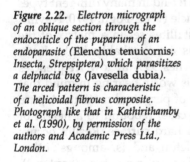
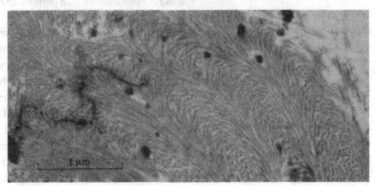

Figure 2.22. Electron micrograph of an oblique section through the endocuticle of the puparium of an endoparasite (Elenchus tenuicornis; Insecta, Strepsiptera) which parasitizes a delphacid bug (Javesella dubia). The arced pattern is characteristic of a helicoidal fibrous composite. Photograph like that in Kathirithamby et al. (1990), by permission of the authors and Academic Press Ltd., London.

vertebrate bone. Being the product of a single layer of cells makes both insect and crustacean cuticles appropriate for experiments on development of extracellular structure. Vertebrate extracellular structures are, by contrast, somewhat more complicated by the presence within them of isolated invasive cells, which interpose their secretions within the original products of an epithelium.

In addition to being found in adults, helicoidal cuticles (as revealed by arced patterns) are found in developmental stages as well. Thus, the embryonic cuticle (serosa) is known to be helicoidal in a grasshopper, *Melanoplus differentialis* (Slifer & Sekhon, 1963),

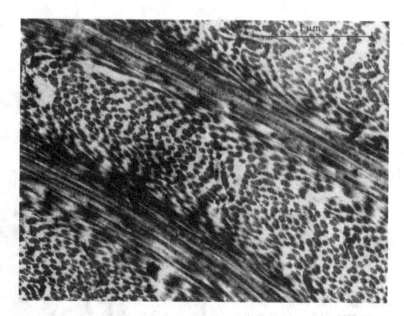

Figure 2.23. Electron micrograph of a section through the cuticle of a crab apodeme (an inwardly facing cuticular projection for muscle insertion). The material is from a shore crab (Carcinus maenas; Crustacea, Malacostraca) and shows well how individual fibrillar components of a helicoidal composite make up an arced pattern. The section is only slightly oblique to the layers of fibrils, which are seen in a sequence of longitudinal, oblique, and transverse sections. Decalcified in EGTA and stained with potassium permanganate and lead citrate. Micrograph by B. M. Luke, reprinted from Neville (1970), by permission of Blackwell Scientific Publications Ltd.

and in a cricket, Acheta domestica (Furneaux, James, & Potter, 1969). A good example of helicoidal layering in a larva is seen in caterpillar cuticle, Calpodes ethlius (Locke, 1960). Published examples of helicoids in a beetle grub, Tenebrio molitor (Neville & Luke, 1969b), in a beetle pupa (Caveney, 1970), and in an adult beetle, Plusiotis resplendens (Caveney, 1971), are representative of countless similar publications.

Helicoidal fibrous composites are found in many different types of cuticle, including rubber-like cuticle (Weis-Fogh, 1970; Neville, 1975a). This consists of chitin crystallites set in a matrix of the protein rubber resilin. Resilin has a remarkably high elastic efficiency (97%), and even approaches 99% at low frequencies (Andersen & Weis-Fogh, 1964). (It is worth bearing in mind that this is close to the impossible 100% of perpetual motion.) A pure piece of resilin would (in an ideal case) rebound to a height of 99 cm if dropped from 1 m onto a hard surface. Resilin occurs in discrete regions of the exoskeleton and is, amongst other functions, involved in flight and flea jumping (see Chapter 3). The occurrence of helicoidally disposed chitin within a rubber with random molecular chain conformation is puzzling.

Helicoids are found also in the intersegmental membranes which link the plates on a locust abdomen (Vincent, 1981). At an appropriate stage of adulthood, and in females only, these membranes become highly extensible (up to fifteen times their original length). This is an adaptation to laying eggs as deeply as possible in the sand or soil. The membranes recover with very slight lasting deformation. The helicoidal parts undergo some reorientation of their chitin crystallites.

There is helicoidal architecture in the foregut cuticle in insects (Smith, 1968) and in the specialized cuticle of the rectum (Noirot & Noirot-Timothée, 1971). It also occurs in the ecdysial membrane cuticle (Neville, 1975a), a thin laminar structure detached from the endocuticle and shed at a moult. The cuticle of sclerites (hard plates), and of the arthrodial membrane in the joints between them, also has helicoidal structure (Neville, 1975a).

Having established the relationship between polarizing and electron microscope images (Fig. 1.20), it then becomes possible to interpret birefringent lamellae as helicoidal cuticle. Some of my students have contributed to survey the occurrences of helicoidal architecture in various locations. The use of very large insects, such as the giant Trinidadian grasshopper, *Tropidacris* (= *Eutropidacris*) *cristata*, and the giant Malaysian stick insect, *Heteropteryx dilatata*, facilitates this task. Helicoids occur in wing veins (Banerjee, 1988a,b), femur, tibia, tarsus, tergite, compound eye lenses, simple ocellar lenses, antennae, mandibles, and ovipositors. The rate and duration of growth of all these structures may be measured by using daily growth layers in the cuticle. In winged insects, the chitin microfibrils in the layers of cuticle are arranged helicoidally if formed at night-time, whereas those formed during the daytime are laid down in parallel.

In addition to the detection of helicoids by arced microfibrillar patterning, they may also be located by the cellular extensions running through the cuticle. These extensions (pore canals) are twisted by the surrounding chitin and generate crescentic patterns in oblique sections (Fig. 3.27B). A species with conveniently large pore canals, the horseshoe crab, *Limulus polyphemus*, has been used to study these patterns (Neville, Thomas, & Zelazny, 1969).

2.3i Brine shrimp eggshells

The eggs of the brine shrimp, *Artemia salina*, have especially thick resistant walls. The eggs, which may be purchased in tubs for use as instant live fish food, can be hatched in warm brine to release nauplius larvae. Electron micrographs of sections through the egg wall show arced patterns characteristic of helicoidal structure (Morris & Afzelius, 1967). These eggs are capable of withstanding cryptobiosis for periods of several years, during temporary drying out of their habitat. Helicoidal structure is a good way to build a strong spherical shell (Table 2.7).

2.3j Fossil arthropod cuticle

The occurrence of helicoidal structure in the cuticle of a fossil crab (*Eryma stricklandi*) from the Jurassic period shows that this type of architecture has been present in arthropods for a long time (Neville & Berg, 1971). Crescentic patterns created by

54 / The occurrence of fibrous composites

Figure 2.24. Electron micrograph of a replica of a freeze-fracture plane through the eggshell of a silk moth (Antheraea polyphemus). The fracture passes through several half-turns of helicoid. Changes in layer direction making up the arced pattern of the helicoid are arrowed. Reprinted from Hamodrakas et al. (1986), by permission of the publishers, Butterworth, Heinemann Ltd.©, and of the authors.

obliquely ground sections of pore canals were compared with those from the cuticle of living crayfish, *Astacus fluviatilis*, and interpreted as part of a helicoidal system.

2.3k Mantis eggcase

The eggs of praying mantids are laid in batches and are surrounded by proteins produced by the left oothecal gland of the female (see Fig. 2.25). The glands are asymmetrical, the right gland producing components concerned with the stabilization of the eggcase proteins by cross-linkage. These proteins originate as liquid crystals with helicoidal texture (Kenchington & Flower, 1969; Neville & Luke, 1971b); they are whipped into a foam around the eggs and then chemically cross-linked. They form a model (with the exception of the foam stage) of the processes thought to be involved in the production of helicoidal insect cuticle.

2.3l Moth eggshells (chorion)

Helicoidal structure in moth eggshells was first interpreted from arced patterns in electron micrographs of the silk moth *Hyalophora cecropia* (Fig. 4.14) by Telfer and Smith (1970) and Smith et al. (1971). Freeze-fracture of chorion shows the individual layers which make up the arcs (Fig. 2.24) that reveal helicoidal structure (Hamodrakas et al., 1986). Helicoids have since been found in the eggshells of many other Lepidoptera (Table 2.4). A very large amount of work has subsequently been carried out on the proteins of moth eggshells and their molecular genetics, espe-

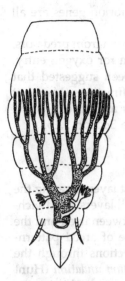

Figure 2.25. Eggcase glands in a praying mantis. Diagram of the body of a female praying mantis showing the location of the large left oothecal gland (stippled). This contains liquid crystalline proteins in a cholesteric (helicoidal) phase; these eventually form the eggcases. This diagram is included because the system forms a developmental model for the formation of helicoidal insect cuticle. The much smaller right oothecal gland, which is involved with the stabilization of the eggcase by cross-linking of protein chains, is also shown (unstippled). The two glands function like the two kinds of tubes of Araldite Epoxy.

cially by Professor F. C. Kafatos and his large group of colleagues at Harvard University (USA). This work has now extended to Athens (Greece) and Heraklyon (Crete). As a result, we now know more about the molecular aspects of moth eggshells than any other helicoidal system. It has become an important paradigm model in attempts to understand which parts of the relevant protein molecules play the vital part in helicoid self-assembly (Hamodrakas, 1984).

Moth eggs are covered by a more or less spherical shell. This consists of an outer layer (chorion), some 60 μm thick in *Hyalophora cecropia*, which is secreted onto the outside of the shell by follicle cells in the ovary of the adult female moth, before ovulation and fertilization take place. Underlying the chorion is another layer (serosa) which is secreted onto the inside of the shell by the developing embryo, after fertilization and nuclear division. There is no chitin in the chorion, and yet it still forms a fibrous composite, in this case of protein microfibrils set in a protein matrix. Using the crude methods available to us at the time (1971), we separated four proteins of distinctly different sizes from *Hyalophora* chorion. These were stabilized in the chorion by disulphide cross-links and hydrogen bonds. By modern methods of separation, huge numbers of proteins (186 in *Antheraea polyphemus*) have subsequently been extracted from silk moth chorion.

Those butterfly and moth chorions which have been investigated have helicoidal structure, ideal for strengthening a spherical shell equally in all directions. I find it unusual that helicoidal structure in the chorion part of insect eggshells appears to be restricted to the single order Lepidoptera. The underlying serosal layer (embryonic cuticle) resembles later cuticles (larvae, pupae, and adults) in consisting of a fibrous composite of chitin rods set in a protein matrix. The serosa is not confined to just a single order of insects.

There is a sequence of layered structures in the chorion. From the inside, these are (1) an inner lamellar layer, (2) a holey layer, (3) an outer lamellar layer, and (4) an oblique layer. All four types of layers are helicoidal, and they have been superbly illustrated by Mazur, Regier, and Kafatos (1982). The distribution of distortions and defects characterizes the four layered regions. The analytical approach of Kafatos and his team was to correlate the appearance of specific groups of proteins with the sequence of formation of the various layers. The proteins were then sequenced, and studies were begun upon their secondary molecular structures.

It is exciting that the chorion of the commercial silk moth (*Bombyx mori*) is also helicoidal (Furneaux & Mackay, 1972). This gives access to the wealth of genetical information about *B. mori* – long a favourite of geneticists. Strongly linked families of genes are involved with chorion formation. The chorion genes are all located on one chromosome (number 2).

The chorion has several functions: (1) protection from predators, (2) protection from desiccation, (3) provision for oxygen entry, and (4) provision for sperm entry. It has been suggested that even the faults in the architecture may have functions (Mazur et al., 1982). For specific details of the molecular aspects of chorion morphogenesis, reference should be made to Section 5.1b.

2.3m Mollusc shell periostracum

The periostracum is the thin brown outermost layer covering the shells of molluscs. It overlies the calcified shell layers, sandwiching the major component (prismatic layer) between itself and the inner layer (mother-of-pearl). The occurrence of arced patterning (Fig. 2.26) in electron micrographs of sections through the periostracum of the gastropod whelk, *Buccinum undatum* (Hunt & Oates, 1970, 1978), indicated helicoidal structure. This is supported by the birefringent laminations seen in sections observed between crossed polarizers (Hunt & Oates, 1978). In *Buccinum* the helicoid angle between consecutive layers is 20–25°. This species has a particularly thick (40 µm) periostracum which may easily be peeled from the shell; this provides a convenient sample for analysis. The periostracum is deposited in the periostracal groove at the edge of the growing shell. The mantle cells in this region have a papillated surface like velvet; it is not surprising, therefore, that the helicoid also shows irregularities. Chemical analysis has shown that the periostracum of *Buccinum* contains a fibrous protein (not collagen) as well as chitin, and subsequent electron microscopy has revealed that both are helicoidally arranged (Hunt & Oates, 1984). It is not known whether the helicoids are interdependent, though this seems likely. (In Chapter 5 it is proposed that the continuous phase in fibrous composites is likely to be the prime organizer.) The protein fibres are arranged in sheets (which rotate to form a helicoid) with a banding pattern, indicating that the components lie in register (Fig. 2.27) without stagger (Hunt & Oates, 1978). The sub-units which make up the protein fibres are thought by Hunt and Oates to be shaped like dumb-bells, 32 nm long and 6.5 nm wide at the globular end.

Helicoidal structure is of restricted occurrence in molluscs. It is found in *Buccinum* and in a terrestrial prosobranch snail (*Pterocyclus latilabrum*), chosen for investigation because of its thick (10 µm) and retained periostracum (Hunt & Oates, 1985). In this species the helicoid takes a 180° half-turn every 0.15–0.5 µm. Helicoids are absent from the periostracum of the winkle (*Littorina littorea*) and the terrestrial snail (*Helix aspersa*), and also from two lamellibranch bivalves examined by Hunt and Oates (1978): *Solen marginatus* and *Arctica islandica*.

2.3 Helicoidal plywoods in animals / 57

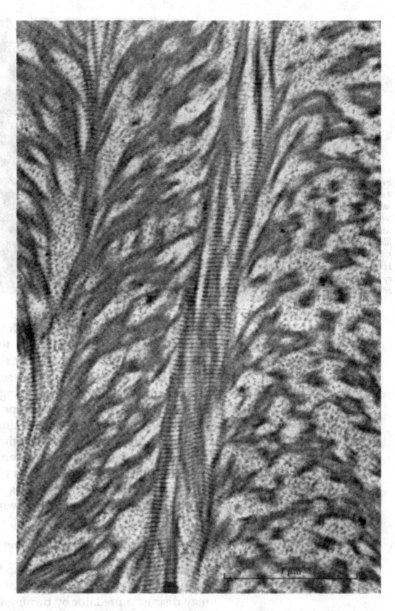

Figure 2.26. Molluscan helicoid. Electron micrograph of a section cut obliquely through the laminated outer protective periostracum (protein) layer of a whelk shell (Buccinum undatum; Mollusca, Gastropoda). The pattern indicates a helicoidal fibrous composite construction. From Hunt and Oates (1978), by permission of the authors and of the Royal Society of London.

Based upon its worn appearance in most marine molluscs, and its retention in their freshwater counterparts, the periostracum is classically regarded as providing chemical protection for the underlying calcareous layers. For example, loss of the periostracum in the freshwater mussel *Anodonta* leads to erosion of the prismatic layer by acid freshwater (Hunt & Oates, 1985). There are, however, two more recent proposals as to periostracal functions.

Extending his ideas from crustacean cuticular calcification, Digby (1968) has proposed that calcification of molluscan shell is

58 / *The occurrence of fibrous composites*

Figure 2.27. Electron micrograph of a section through the periostracum which forms the outer layering of a whelk shell (Buccinum undatum; Mollusca, Gastropoda). The protein fibrils are regularly banded and lie in register; they are not collagen. From Hunt and Oates (1978), by permission of the authors and of the Royal Society of London.

under electrochemical control. The periostracum is cross-linked (tanned) by aromatic molecules which have an alternation of single and double bonds, so that they are able to transmit electricity as a semi-conductor. This sets up an electrochemical potential with calcium carbonate being deposited in electro-positive regions (to the inner side of the periostracum). Hunt and Oates (1978) make the thought-provoking suggestion that the periostracum forms a barrier between the underlying mantle cells and the seawater, making it easier to control the extracellular environment in which calcification takes place.

It is possible that the eggcases of whelks contain helicoidal fibres (Hunt, 1971), but further work is desirable.

2.3n Sea cucumber defensive secretion

Sea cucumbers are a class of echinoderms (Holothuroidea) which have unusual defence mechanisms. If molested, some species may distract a predator by turning their body viscera inside out. Others, such as *Holothuria forskali* (a 20-cm-long species from the Atlantic and the English Channel), may eject sticky white threads through the anus – hence their other name of 'cotton-spinners'. The secretion is produced by Cuvierian glands and contains collagen microfibrils in an analogue of a mobile liquid crystalline phase (Dlugosz, Gathercole, & Keller, 1979). Electron micrographs of sections of the fixed secretion show arced patterning characteristic of helicoidal texture (Fig. 2.28). The collagen fibrils change orientation during expulsion from an analogue of a convoluted cholesteric liquid crystal to a predominantly parallel fibre.

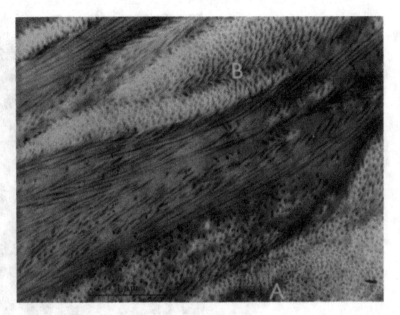

Figure 2.28. Electron micrograph of a section through the sticky contents of a Cuvierian tube from a sea cucumber (Holothuria forskali; Echinodermata, Holothuroidea). These contents are in the form of a liquid crystal, because the patterns made by the collagen fibrils indicate a cholesteric phase (at B), as well as other phases elsewhere (e.g. parallel at A). From Dlugosz et al. (1979), reprinted with permission from Maxwell Pergamon Macmillan plc, Oxford, and from the author.

The fibrils have been identified as collagen by X-ray diffraction (Bailey et al., 1982). They have the 68-nm banded repeat typical of type I collagen, with twelve subsidiary bands visible in the electron microscope. The system contains forty times more glycosaminoglycans than rat tail collagen, and these interact with the collagen at the 68-nm periodicity. Bailey et al. (1982) suggest that this large amount of glycosaminoglycans forms a sheath around the collagen fibrils which inhibits cross-links between them and hence reduces their tensile strength. This system forms a useful model of collagen morphogenesis, to compare with other systems like bone and cornea.

2.3o Tunicate (sea-squirt) test

Sea squirts (Tunicata) are invertebrates, with sufficient characters in common with humans as to be catalogued in the same phylum (Chordata). The ectodermal epidermis of sea squirts secretes onto its outer face an extracellular test which contains cellulose fibres. (Formerly known as tunicin, it is now accepted from X-ray diffraction evidence that the fibres of sea-squirt test are identical to plant cellulose.) The test contains a few cells which are isolated between the extracellular fibres (Fig. 2.29), a situation found also in vertebrate skeletal systems. The fibres in some species are arranged somewhat erratically to form a vesicular test. However, in *Halocynthia papillosa* the cellulose forms a helicoid. We have here an animal with helicoidal cellulose. Electron microscopy shows that the cellulose is in the form of wide

60 / *The occurrence of fibrous composites*

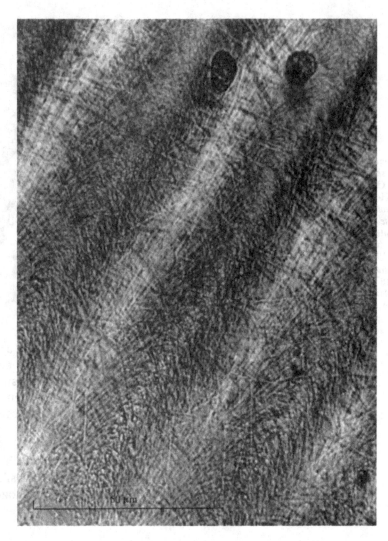

Figure 2.29. Photomicrograph using polarized light of a section through the test of a Mediterranean sea squirt (Halocynthia papillosa; Chordata, Tunicata). The test is a helicoid constructed with large fibrils of cellulose. A characteristic arced pattern is seen, secreted by the underlying epidermis. There are also a few fibres which cross the main helicoidal laminate, secreted perhaps by the isolated cells which are found inside the helicoidal test; two can be seen in the micrograph.

bundles, themselves constructed from thinner crystallites. Because of the wide-diameter fibres and long pitch (up to 100 μm per 360° rotation) of the helicoid, this system is well suited for light microscopical demonstrations involving helicoids. For example, the optical procession diagram drawn in Figure 3.19 was based upon observations using *Halocynthia*. The large fibres are also convenient for scanning electron microscopy and have been used (Gubb, 1975) to show that arced patterns in sections of helicoids are made up from short lengths of fibres which are themselves straight (Figs. 1.13 and 1.14). Gubb's paper gives full evidence for the interpretation of *Halocynthia* test as helicoidal. The time seems right for a developmental study of tunicate test, using the bright scarlet species *Halocynthia papillosa*. This is a Mediterra-

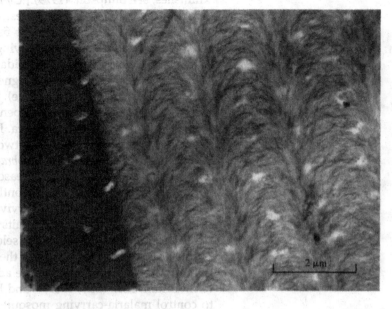

Figure 2.30. Electron micrograph of a section cut obliquely through an eggshell of a codfish (Gadus morrhua). The microfibrils form an arced pattern characteristic of a helicoidal system. The non-helicoidal electron-dense external cortex is secreted first, with the helicoidal inner cortex added later to its inside surface. From Grierson and Neville (1981), by permission of Churchill Livingstone Journals, Edinburgh, U.K.

nean species, so a laboratory close to its source would seem appropriate.

2.3p Fish eggshells

Several previously published papers recorded arced patterns in electron micrographs of sections through fish eggshells, but these were not interpreted as helicoids until a decade ago (Grierson & Neville, 1981). We studied helicoidal eggshells in cod (*Gadus morrhua*), plaice (*Pleuronectes platessa*), and trout (*Salmo gairdneri*). A list of helicoidal architecture in fish eggshells is given in Table 2.6. Both the outer and inner layers of fish eggshells are secreted as extracellular products through the oocyte cell membrane. The outer layer (primary chorion or cortex radiata externa) is usually about 3 μm thick, whereas the inner layer (secondary chorion or cortex radiata interna) may reach a thickness of 55 μm in the eggs of cod (*Gadus morrhua*). The proteins of the latter show characteristic arced patterns (Fig. 2.30) in electron micrographs; they share remarkable convergent evolution with the chorion of moth eggs. The arced patterns in fish eggshell originate in the helicoidal microfibrillar system, whereas the crescentic pattern arises from sections of the radial canals. These traverse the chorion and are forced into flattened and twisted ribbons (Fig. 3.27A) by the surrounding helicoid. They are convergently evolved with the shapes of pore canals in insect cuticles (Fig. 3.27B).

One of the species known to have arced patterns is *Cynolebias belotti* (Sterba & Müller, 1962). Interestingly, this is an annual killifish – an egg-laying tooth carp. [For a general review of killifishes, see Simpson (1979).] *Cynolebias* belongs to a group of fish which inhabit temporary freshwater pools. These may dry out completely, killing the adult fish. The eggs are laid in the mud and are able to survive drying out. They hatch when the next rainy season arrives. Helicoidal structure is often found in the shells of eggs and spores designed to resist spells of drought. Among killifish (Cyprinodontidae), the habit of laying drought-resistant eggs has evolved independently in tropical and subtropical Africa and South America. In South America, *Cynolebias* grows so rapidly that it fits in two wet seasons and two dry seasons per year. In Africa, *Nothobranchius* goes through a single wet season and a single dry season per year. *Cynolebias* can survive a drought of up to ten months' duration, whereas *Nothobranchius* has been known to survive a drought lasting for two years. An interesting project awaits someone who could investigate chorion helicoids in a wide selection of these annual fishes. Enthusiasts exchange the eggs of these species by post (enclosed in damp peat). They hatch on the addition of water – hence the term 'instant fish' (Bay, 1965) – and have been used in rice fields to control malaria-carrying mosquitos.

Fish eggshells function in the control of gaseous diffusion, physical protection, and provision for sperm entry. The observation by Anderson (1967) that the inner chorion of fish eggshell changes its appearance from reticulated at an early stage to multilaminated at maturity is important. We have suggested that this indicates that the helicoidal chorion self-assembles (Grierson & Neville, 1981).

2.3q Human bone

Bone provides the internal skeletal support in vertebrates. It consists of a calcified fibrous composite, first recognized as such in an important paper by Currey (1962). In mammals, bone is arranged in concentric cylinders called osteons. Collagen fibres occupy about half of the volume of bone and are arranged in precise directions. Recently, in an important publication, Giraud-Guille (1988) has obtained convincing electron micrographs of collagen fibre orientations in sections of decalcified human bone osteons (Fig. 2.31). In some regions, flat (planar) orthogonal layers grade into co-axial orthogonal osteon cylinders by rolling up (Fig. 2.32A). Likewise, regions of flat planar helicoidal layers grade into co-axial helicoidal osteon cylinders (Fig. 2.32B). Occasionally, a helicoidal osteon may show transition into an orthogonal osteon.

Bundles of osteons are cemented together to form a haversian system. The bone cells (osteocytes) are located within the bone structure and are surrounded by spaces (lacunae). The spaces are

2.3 Helicoidal plywoods in animals / 63

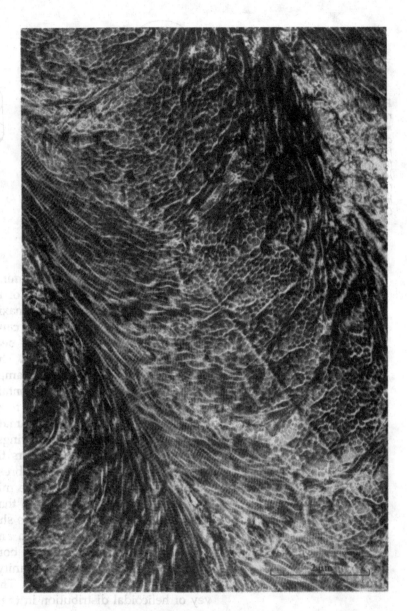

Figure 2.31. *Helicoids in humans. An important electron micrograph of human bone. A femur of a thirty-five-year-old human (Homo sapiens) was decalcified with EDTA and cut obliquely to the shaft. The striated collagen fibrils make up an arced pattern, revealing their helicoidal arrangement in compact bone osteons. From Giraud-Guille (1988), by permission of the author and of Springer-Verlag, New York.*

interconnected by small (0.2-μm-diameter) passages (canaliculae) containing processes of the osteocytes, which form cell–cell junctions. Bone differs from cartilage in this respect, and also because it is supplied with blood vessels running axially along the osteon cylinders and giving off lateral branches. The blood vessels contact the osteocyte processes.

Bone collagen is set in a matrix of polysaccharides. It calcifies with hydroxyapatite, the nucleation sites of the mineral crystallites being controlled by the collagen fibrils. Developing bone stiffens

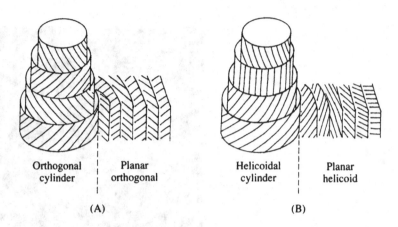

Figure 2.32. Three-dimensional representation of osteons from mammalian bone. Each osteon is a cylinder of coaxial layers of collagen fibres. Two types are shown. (A) A flat (planar) orthogonal region rolls into an orthogonal cylinder. (B) A flat (planar) helicoidal region rolls into a helicoidal cylinder. Redrawn from Giraud-Guille (1988).

gradually at first by increasing mineral content, and then more suddenly by end-to-end fusion of mineral crystallites. There is an optimal mineral content for maximum toughness, defined by the amount of energy a material can absorb before it fractures in response to a blow. Bones have become adapted to match this optimal value of mineralization. There exists a wide variety of bone types, including extreme examples such as the mature bones of teleost (bony) fishes, which contain neither osteocyte cells nor a blood system. For this reason they grow only at the surfaces, instead of from within like mammalian bone; they can therefore produce zoned annual growth rings.

Future research on bone offers the challenge of finding out how the directions of collagen fibres are controlled; they in turn influence the orientations of the mineral crystals which stiffen developing bone. The revelation that bone contains both orthogonal and helicoidal architecture should have repercussions on attempts to estimate its strength and stiffness. The findings of Giraud-Guille (1988) in human bone should serve to call the attention of the medical community to helicoidal systems; this augurs well for future research. There is also scope for a survey of helicoidal distribution in the bones of other mammalian species.

2.4 Helicoidal plywoods in plant cell walls

Helicoidal structure was discovered in animal material. It was only a matter of time before the principle was applied to the structural systems of plants. A helicoidal explanation was independently proposed for the walls of three types of plants by three different groups: for mung bean hypocotyl, *Phaseolus*, by Roland, Vian, and Reis (1977); for the eggs of the seaweed *Pelvetia* by Peng and Jaffe (1976); and for the oospores of the stonewort alga *Chara* by Neville et al. (1976). Two of these three examples are algae; the lower plants serve as a morphogenetical model for higher plants.

Each plant cell secretes around itself a thick extracellular box known as the cell wall. This has important consequences for the way of life of plants. The skeletal support depends partly upon the boxes being glued together. This static support is aided by the turgor pressure of the fluid in the cytoplasm of each cell. The boxes limit plant mobility, so that they are sessile and forgo any equivalent of the behaviour found in animals. The boxes prevent the development of a contractile circulatory system, leading to an alternative: pumping of water by transpiration. The cell wall may form the boundary of the plant, in which case that face which is external may be covered with a waterproofed cuticle; alternatively, it may be surrounded on all faces by other cells.

The wall consists of a fibrous cellulose composite, set in a chemically complex matrix of hemicelluloses and pectin, often hardened with lignin. It has emerged that the directions of cellulose microfibrils are under precise control. Growth in cell volume coincides with the deposition of the *primary* (external) layers of its wall. This is therefore subject to growth forces, which usually reorient the cellulose microfibrils within it. When growth in cell dimensions is complete, a *secondary* wall is deposited to the inside of the primary wall. The cellulose orientations of the secondary wall are therefore not modified by growth forces and so can provide information on undistorted cellulose architecture. There are similar relationships in microfibril positioning between primary versus secondary plant cell walls, and exocuticle versus endocuticle in insects. Arced patterns due to helicoids are found in both primary and secondary cell walls. In primary walls the pattern is often transitory; growth-induced changes in pattern are known as dissipation (Roland et al., 1982).

Plant cell walls are involved in many functions. Their primary task is to provide skeletal support. They play a dominant role in plant cell growth and morphogenesis, in disease resistance, in cell recognition and signalling, in digestibility, herbivore nutrition, and human dietary fibre, in plant decay, and as a newly discovered source of oligosaccharin hormones. Further details of these topics have been discussed by Fry (1988). An appreciation of cellulose orientation is also relevant to the commercial application of fibres in paper pulps, ropes, textiles, and timbers.

The literature is quite rich in uninterpreted electron micrographs of plant cell walls showing arced patterning. These are listed, along with mainstream sources of information on helicoids, in Tables 2.8–2.11, which cover algae, lower plants, gymnosperms, monocotyledons, and dicotyledons.

Table 2.8. *Algae for which electron micrographs of cell walls show arced patterns which indicate microfibrils oriented as helicoids. Based mainly on Neville (1988c).*

Group	Species	Structure	Reference
Chlorophyta (green algae)	*Cylindrocapsa geminella*	Zoospore	Hoffman & Hofmann (1975)
	Spongomorpha arcta	Cell wall	Hanic & Craigie (1969)
	Pithophora oedogonia	Cell wall	Pearlmütter & Lembi (1978, 1980)
	Boergesenia forbesii	Aplanospore	Mizuta & Wada (1982)
	Boodlea coacta	Thallus wall, Protoplast wall	Mizuta & Wada (1981) Mizuta (1985)
	Chamaedoris orientalis	Thallus wall	Mizuta et al. (1989)
	Oocystis solitaria	MBC-treated cell wall	Robinson & Herzog (1977)
	Closterium ehrenbergii	Zygote cell wall	Noguchi & Ueda (1985)
Phaeophyta (brown seaweeds)	*Pelvetia fastigiata*	Egg cell wall	Peng & Jaffe (1976)
	Fucus serratus	Zygote	Callow et al. (1978)
Bacillariophyta (diatoms)	*Odontella laevis*	Valve wall	Crawford (personal communication)
Charophyta (stoneworts)	*Chara vulgaris*	Oospore wall	Baynes (1972), Neville et al. (1976)
	Chara corallina	Cell wall	Neville & Levy (1984)
	Chara delicatula	Cell wall	Leitch (1986)
	Chara hispida	Oospore wall	Leitch (1986, 1989)
	Nitella opaca	Inner cell wall	Probine & Barber (1966), Neville & Levy (1984, 1985)
		Oosporangium	Leitch (1986)
	Nitella translucens	Cell wall	Levy (1986)
	Lamprothamnion papulosum	Oosporangium, spiral wall	Leitch (1986, 1989)

2.4a Blue-green algae (cyanobacteria)

The aquatic blue-green algae are primitive and represent the source by endosymbiotic capture of the chloroplasts of higher plants. It is therefore most interesting that the mucilage secreted into the surrounding water by one species (*Rivularia atra*) shows arced patterning derived from helicoidal structure, when fixed, sectioned, and examined in the electron microscope. This example is illustrated by Bouligand (1978b) with a micrograph by J. C. Thomas. The secretion, which is liquid crystalline, indicates the probable early origins of helicoidal systems. Although not a plant cell wall, this example is included here because of its evolutionary interest.

2.4b Algae

There is remarkable diversity in cell wall microfibrillar architecture among the algae (Neville, 1988c). Helicoidal structure occurs in the zygote cell walls of seaweeds such as *Fucus serratus*

Table 2.9. *Lower plants (excluding algae) for which electron micrographs of cell walls show arced patterns which indicate microfibrils oriented as helicoids*

Group	Species	Structure	Reference
Fungi	*Synchytrium mercurialis*	Sporocyst	Meyer & Michler (1976)
	Allomyces neo-moniliformis	Sporangium	Skucas (1967)
	Endogone sp.	Spore wall	Mosse (1970)
	Glomus epigaeum	Spore wall	Bonfante-Fasolo (1983)
	Glomus versiforme	Spore wall	Bonfante-Fasolo et al. (1986)
	Glomus macrocarpum	Spore wall	Bonfante-Fasolo (personal communication)
Bryophyta (mosses)	*Rhacopilum tomentosum*	Peristome wall	Schnepf et al. (1978)
	Rhytidiadelphus squarrosus	Leaf cell walls	Mabb (1989)
		Stem cell walls	Mabb (1989)
Sphenopsida (horsetails)	*Equisetum hyemale*	Inner wall, root hairs	Sassen et al. (1981) Emons (1982)
	Equisetum fluviatile	Root hairs	Emons & van Maaren (1987)
	Equisetum palustre	Root hairs	Emons & van Maaren (1987)
	Equisetum variegatum	Root hairs	Emons & van Maaren (1987)
Filicopsida (ferns)	*Ceratopteris thalictroides* (water fern)	Root hairs	Sassen et al. (1981)
Lichens	*Peltigera canina*	Fungal wall	Boissière (1987)

Table 2.10. *Coniferous gymnosperms for which electron micrographs of cell walls show arced patterns which indicate microfibrils oriented as helicoids*

Species	Structure	Reference
Pinus strobus (Weymouth pine)	Sieve wall	Chafe & Doohan (1972)
Pinus sylvestris (Scots pine)	Needle tracheids	Liese (1965)
Cryptomeria japonica (Japanese cedar)	Xylem parenchyma	Chafe (1974)
Picea abies (Norway spruce)	Tracheid	Parameswaran & Liese (1982)
Pseudotsuga menziesii (Douglas fir)	Bark sclerotic fibres	Parameswaran & Liese (1981)

(Callow, Coughlan, & Evans, 1978), in the egg cell walls of *Pelvetia fastigiata* (Peng & Jaffe, 1976), and in the cell walls of many other species (Table 2.8). There is also a helicoidal layer in the cell wall of a diatom, *Odontella laevis* (Dr. R. M. Crawford, personal communication). We have scanned the phycological literature for arced patterns in electron micrographs, which indicate helicoids. There are undoubtedly many we have missed. Helicoids occur in Chlorophyceae, Phaeophyceae, and Bacillariophyceae. There are examples in both unicellular and multicellular species, in spores, zygotes, and thalli, and from marine as well as freshwater species. The helicoids may be of cellulose or sometimes of chitin (e.g. in *Pithophora*).

Table 2.11. *Flowering plants (angiosperms) for which electron micrographs of cell walls show arced patterns which indicate microfibrils oriented as helicoids*

Species	Structure	Reference
Monocotyledons		
Hydrilla verticillata	Epidermis	Pendland (1979)
Phyllostachys edulis (bamboo)	Culm parenchyma	Parameswaran & Liese (1975, 1980)
Cephalostachyum pergracile (bamboo)	Culm parenchyma	Parameswaran & Liese (1980)
Oxytenanthera abyssinica (bamboo)	Culm parenchyma	Parameswaran & Liese (1980, 1981)
Hordeum vulgare (barley)	Epidermal wall	Sargent (1978)
Zea mays (maize)	Coleoptile	Satiat-Jeunemaître (1981)
Lolium temulentum (darnel)	Sclerenchyma fibres	Juniper et al. (1981)
Echinochloa colonum	Leaf-base collenchyma	Parker (1979)
Juncus effusus (soft rush)	Pith stellate cells	Roland (1981)
Limnobium stoloniferum (= *Trianea bogotensis*)	Root hairs	Pluymaekers (1980, 1982), Sassen et al. (1981)
Zostera capensis (marine grass, wrack)	Leaf epidermis	Barnabus et al. (1977)
Thalassia testudinum (marine grass)	Leaf epidermis	Jagels (1973), Benedict & Scott (1976)
Thalassia hemprichii (marine grass)	Leaf epidermis	Doohan & Newcomb (1976)
Halophila ovalis (marine grass)	Leaf epidermis	Birch (1974)
Cymodocea rotundata (marine grass)	Leaf epidermis	Doohan & Newcomb (1976)
Cymodocea serrula (marine grass)	Leaf epidermis	Doohan & Newcomb (1976)
Butomus umbeliatus (flowering rush)	Root hairs	Emons & van Maaren (1987)
Stratiotes aloides (water soldier)	Root hairs	Emons & van Maaren (1987)
Hydrocharis morsus-ranae (frogbit)	Root hairs	Emons & van Maaren (1987)
Ruppia maritima (tassel pondweed)	Root hairs	Emons & van Maaren (1987)
Zebrina purpusii	Root hairs	Emons & van Maaren (1987)
Phragmites australis (reed)	Root hairs	Emons & van Maaren (1987)
Cyperus papyrus (papyrus paper reed)	Aerenchyma	Mosiniak (1986), Mosiniak & Roland (1986)
Stipa tenacissima (desert alpha grass)		Harche (1986)
Dicotyledons		
Phaseolus vulgaris (French bean)	Pod epidermis	Tapp (1987)
Phaseolus aureus (mung bean) (= *Vigna radiata*)	Hypocotyl cell wall	Roland et al. (1977, 1982), Reis et al. (1978), Vian (1978), Reis (1981), Roland (1981)
Vigna angularis (Azuki bean)	Epidermis	Takeda & Shibaoka (1981)
Pisum sativum (pea)	Epidermis, cortex, pith	Appelbaum & Burg (1971)
Lactuca sativa (lettuce)	Hypocotyl	Sawhney & Srivastava (1975)
Apium graveolens (celery)	Petiole collenchyma	Chafe & Wardrop (1972), Cox & Juniper (1973), Roland & Vian (1979), Roland (1981)
Cucurbita pepo (marrow)	Petiole parenchyma	Deshpande (1976)
Rumex conglomeratus (sharp dock)	Collenchyma	Chafe (1970), Chafe & Wardrop (1972)
Petasites fragrans (winter heliotrope)	Collenchyma	Chafe (1970)
Helianthus annuus (sunflower)	Epidermis	Rajaei (1980)
Aristolochia clematitis (birthwort)	Stem sclerenchyma	Roland et al. (1987)
Tilia platyphyllos (broad-leaved lime)	Wood tracheids, fibres, parenchyma	Roland (1981), Roland & Mosiniak (1983)
Fagus sylvaticus (beech)	Bark sclereids	Parameswaran & Sinner (1979), Parameswaran (1975)
Populus tremuloides (aspen)	Xylem parenchyma	Chafe & Chauret (1974)
Populus eurameniana	Sclereids, secondary phloem	Nanko et al. (1978)
Entandophragma candollei (mahogany)	Bark sclereids	Parameswaran (1975)

Table 2.11. (cont.)

Species	Structure	Reference
Cydonia oblonga (quince)	Pigment cell walls	Abeysekera & Willison (1988)
	Seed epidermal mucilage	Willison & Abeysekera (1988)
Chaenomeles speciosus (quince)	Seed epidermal walls	Neville & Martin (unpublished)
Corylus avellana (hazel)	Nut stony pericarp	Roland et al. (1989)
Pyrus malus (pear)	Stone cells in fruit flesh	Vian & Roland (1987), Roland et al. (1987)
Prunus sativum	Endocarp stone cells	Roland et al. (1987)
Prunus domestica (prune)	Fruit endocarp	Reis et al. (1992)
Prunus cerasus (cherry)	Endocarp sclereids	Reis et al. (1992)
Juglans regia (walnut)	Endocarp sclereids	Reis et al. (1992)
Olea europea (olive)	Fruit endocarp, mesocarp sclereids	Ougherram (1989)

2.4c Stoneworts (Charophyta)

In evolutionary terms, stoneworts are situated more or less between the green algae and the mosses (Bryophyta): they are all aquatic and attach to the substrate by rhizoid roots. They are found in freshwater lakes and drainage ditches, and there are also some in brackish or even quite salty water. The female reproductive bodies (oosporangia) have a calcified wall from which the name stoneworts derives. The spores fall to the bed of the habitat to overwinter.

Stoneworts have proved popular for use in studies on plant cell wall morphogenesis and mechanical properties because of their amazingly long internodal cells, which may measure up to 30 cm in some species. These syncytial cells are the result of the fusion of up to 1,000 normal cells; they thus come to possess 1,000 nuclei, and so provide sufficient genetic material to control a giant cell. In *Nitella*, a single cell forms each internode, whereas in *Chara* the axial internodal cell is surrounded by small cortical cells. Because they are aquatic, it is possible to grow stoneworts in artificial solutions, to which may be added morphogenetic chemicals, such as colchicine.

Electron micrographs of stonewort wall sections by Baynes (1972) and by Probine and Barber (1966) show arced patterning, leading us to suspect that they may have helicoidal structure. This was confirmed for *Nitella* internode wall (Fig. 2.33) by Neville and Levy (1984) and for *Chara* oosporangium wall by Neville et al. (1976).

The oosporangia are the most complex in algae or their relatives. Five spiral cells coil anti-clockwise around the egg cell and then mineralize, providing it with protection. The wall of the oosporangium is complicated, with eight different kinds of

70 / *The occurrence of fibrous composites*

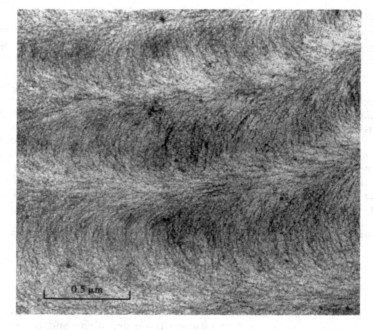

Figure 2.33. Helicoidal plywood in plants. Electron micrograph of an oblique section through a cell wall of a long internodal cell from a freshwater stonewort alga (Nitella opaca; Charophyta). The ranks of arced patterning indicate helicoidal construction. Specimen stained with uranyl acetate in the block, followed by lead citrate on the section. Micrograph by Dr. S. Levy, from Neville and Levy (1984), by permission of Springer-Verlag, Heidelberg.

layers (Leitch, 1986), including the helicoidal layer. The calcified oosporangia are stable enough to fossilize and date back to the late Silurian period. Known as *gyrogonites*, they are used in precise stratigraphical dating of rock strata by geologists.

2.4d Fungi and lichens

Walls with helicoidal chitin microfibrillar architecture (Fig. 1.15) are found in the spores of some fungi, such as *Glomus epigaeum* (Bonfante-Fasolo, 1983). This species forms a mycorhizal association with the roots of the maidenhair tree (*Ginkgo biloba*), a tree whose only other relatives are found as fossil leaves in Jurassic deposits. This fungus is unusual in that it has an aerial rather than subterranean fruiting body. As far as I am aware, no examples of helicoids are known in the cell walls of fungal *hyphae*; perhaps this is because helicoid formation is incompatible with the tip growth characteristic of fungal hyphae.

I am grateful to Dr. D. G. Brown for bringing to my attention a paper on lichen walls by Boissière (1987), who has independently observed and interpreted a quarter-turn of helicoid in the cell walls of the lichens *Peltigera canina* and *Lasallia pustulata*. Technically this is in a fungal hyphal wall, as this forms the outer protective part of the lichen symbiosis between fungus and alga. *Peltigera canina* has a thick flexible thallus with long rhizoids underneath. The microfibrils forming the helicoid are chitin.

2.4e Mosses and liverworts (Bryophyta)

Helicoids in mosses were interpreted by Neville and Levy (1984) from arced patterning in an electron micrograph of a section through the cell walls of the peristome (pepper pot) of *Rhacopilum tomentosum* (Schnepf, Stein, & Deichgräber, 1978). A further uninterpreted example of moss cell wall arced patterns is found in an electron micrograph by Mabb (1989), for leaf and stem cell walls of *Rhytidiadelphus squarrosus*. This shows five half-turns of helicoid. More work is needed on the occurrence of helicoids in moss cell walls. I am grateful to Professor G. Duckett for the observation that most moss materials are chosen for nutrient experiments and hence tend to have thin cell walls. These are less likely to be helicoidally multilamellate than are cells with thicker walls. As far as I am aware there are no published records of helicoids in liverworts, and these could also repay study in this context.

2.4f Horsetails and ferns (Pterydophyta)

Helicoidal wall structure has been deduced from arced patterning in sections of root hairs in several species (Table 2.9) of horsetails, *Equisetum* (Sassen et al., 1981; Emons, 1982; Emons & van Maaren, 1987). The Dutch group (Sassen et al., 1981) has also located helicoids in root hairs of a Sumatran water fern (*Ceratopteris thalictroides*), but so far there does not seem to be any record for land ferns. Ferns and their relatives seem promising material for further investigation. There are signs of zonation in fern spores (Steve Martin, personal communication), and the quillwort (*Isoetes*), a fern found in freshwater lakes, could repay investigation, as it produces two sorts of spores – one small and dust-like, and the other large enough to be seen by the naked eye. The large spores, which resemble insect eggs, seem promising material. Another question is whether there are any helicoids in club mosses (*Lycopodium*).

2.4g Angiosperms (monocotyledons)

A considerable literature has accumulated on helicoidal structures in monocotyledons, some of it reinterpreted from existing published micrographs. The parenchyma cell walls of various bamboos were worked on by Parameswaran and Liese (1980, 1981), while one of the favourite materials used by Satiat-Jeunemaître (1981, 1984) was the coleoptile of maize corn (*Zea mays*). The coleoptile is the first stem which carries aloft the single cotyledon leaf after seed germination.

Rushes have leafless stems which contain pith. The pith is composed of star-shaped (stellate) cells, glued together to leave large spaces which permit easy diffusion of oxygen down to the

72 / *The occurrence of fibrous composites*

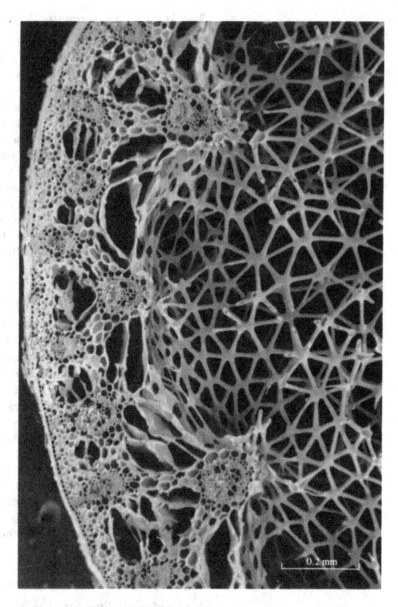

Figure 2.34. Scanning electron micrograph of a section through a stem of a rush (Juncus effusus). The pith which fills the middle of the stem is made of a geodetic construction of tubular stellate cells, whose walls are known from arced patterning in transmission electron microscopy to be helicoidal (Roland, 1981). Micrograph from work by S. Martin and A. C. Neville.

roots; this allows rushes to grow in stagnant conditions. Such pith is known as aerenchyma. The stems of *Juncus effusus* are deep green, with fine grooves. A low-power scanning electron micrograph of this species is presented in Figure 2.34. Roland (1981) showed that rush aerenchyma cells have helicoidal walls, and the same features have been shown (Mosiniak & Roland, 1986) in the papyrus paper reed (*Cyperus papyrus*). This is the material used in making ancient Egyptian paper. Further details of the surprising mechanical adaptations of rush pith are given in Section 3.2.

The eel grasses (e.g. *Zostera*) are the only *flowering* plants which are rooted into the substrate so as to grow totally submerged in the sea. They are found on muddy shores at or below low-water mark and are exposed only at the lowest spring tides. Arced patterns in the leaf epidermis walls of marine grasses have been described in several papers in the literature (Table 2.11) and have been interpreted as helicoids (Neville & Levy, 1984). Table 2.11 also lists several other plants (many of them semi-aquatic) which have helicoidal walls in their root hairs (Emons & van Maaren, 1987).

2.4h Angiosperms (dicotyledons)

The bean sprouts popular in Chinese restaurants represent the seedling stage following germination of the mung bean, *Phaseolus aureus*. The epidermal cells of the hypocotyl (where the stem bends like a bishop's crosier) have been used extensively in experiments by Professor J.-C. Roland and co-workers in Paris (Table 2.11). The outermost layer of cells has helicoidal wall structure, which later dissipates as the epidermal cells elongate.

In the closely related dwarf bean (*Phaseolus vulgaris*) the whole pod is used for human consumption. The pod epidermal cells have helicoidal structure (Tapp, 1987); as the entire pod is eaten, this forms part of the human dietary fibre intake.

Many other dicotyledonous plants are now known to have helicoidal cell walls (Table 2.11). They include the epidermis of the hypocotyl of sunflowers, *Helianthus annuus*, studied by Rajaei (1980), and the collenchyma of the winter heliotrope, *Petasites fragrans*, which was originally interpreted as polylamellate by Chafe (1970) and reinterpreted by Neville and Levy (1984).

A significant recent discovery was the finding of helicoidal texture (Fig. 4.16) in the mucilage released by the epidermis of quince seeds (*Cydonia oblonga*) (Abeysekera & Willison, 1987, 1988). This mucilage is in a cholesteric liquid crystalline phase and as such may serve as a model of plant cell wall self-assembly. This is discussed in Section 4.3f. We have found helicoidal structure in the seed epidermal walls of a closely related quince, *Chaenomeles japonica*.

2.4i Trees (angiosperms and gymnosperms)

It has emerged during the past decade that several parts of trees have cells with helicoidal walls (Tables 2.10 and 2.11). The *bark* of several trees contains sclereid fibres with helicoidal walls [e.g. beech (*Fagus sylvaticus*), mahogany (*Entandophragma candollei*), and Douglas fir (*Pseudotsuga menziesii*)]; see Parameswaran (1975) and Parameswaran and Liese (1981). This seems to make mechanical sense, as the bark forms the outer protecting layer of a tree.

Wood is a natural plywood. Several types of wood cells include helicoidal structure, such as tracheids, fibres, parenchyma, sclereids, and phloem (Roland, 1981; Roland & Mosiniak, 1983) (Table 2.11). Helicoidal structure is a factor which needs to be taken into account when analysing the strength of wood. Further details of wood tracheid ultrastructure are given in Section 2.5, as it is an example of pseudo-orthogonal structure. It would be interesting to know if helicoidal cell walls occur in cycads – a group of living fossil trees.

2.4j Fruits

Fruit sclereids or *stone cells* form the gritty texture noticed when one eats the flesh of a pear (*Pyrus malus*). They prove to have astonishingly thick helicoidal cell walls (Fig. 2.35), with the characteristic arced patterning visible at higher magnification (Fig. 2.36). These superb micrographs were obtained by Vian and Roland (1987) and show up to fifty complete turns of helicoid. The function of such heavily reinforced structure remains a mystery.

In the nut shells (pericarp) of hazel trees (*Corylus avellana*), electron microscopy shows that thick helicoidal cell walls almost totally occlude the cytoplasm (Roland et al., 1989). Neighbouring cells are in contact with each other by lined-up pits with plasmodesmata (Fig. 2.37). The combination of thick helicoidal walls and lignification results in extremely good protection for the seeds during dispersal.

2.4k Types of plant cells with helicoidal walls

Helicoids have been found in a wide variety of cell types. Examples are illustrated in Figure 2.38. They may occur in both primary and secondary walls, although in primary walls they may be transient. They are also found in the spherical walls of spores, oospores, and zoospores.

Parenchyma cells function as aerenchyma (Fig. 2.38A) in the pith of a rush stem (*Juncus effusus*). The cells are stellate (star-shaped) and are fused to their neighbours. Electron microscopy has shown that these walls are helicoidal (Roland, 1981). It is suggested that these cells may prove important in future chemical studies because of the ease with which they can be isolated from other types of cells.

A parallel bundle of sclerenchyma *fibres* (Fig. 2.38B) is drawn cut across to show their thick secondary cell walls. Such walls are known to be helicoidally laminated.

Collenchyma cells (Fig. 2.38C) also have thick secondary walls, especially in the angles. In cases which have been analyzed carefully these are known to be helicoidal.

Figure 2.35. A low-power electron micrograph of a section through the walls of two stone cells from the fruit of a pear (Pyrus communis). Stone cells form the grit noticed when eating a pear. Stone cells have remarkable helicoidal walls, with up to 100 half-turns (180°) – equivalent to 18,000° of layer rotation. Arced patterns can be resolved only at higher magnification, as in Figure 2.36. The cell lumen is almost completely occluded, but the cells remain in contact with their neighbours via pits. The function of these cells is unknown. From Vian and Roland (1987), by permission of the authors and of Cambridge University Press.

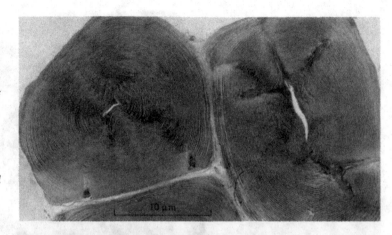

Figure 2.36. As for Figure 2.35, but at higher magnification. The extensive arced patterning in the helicoidal cell wall is well shown. From Vian and Roland (1987), by permission of the authors and of Cambridge University Press.

Sclerenchyma (*sclereid*) cells (Fig. 2.38D) also have helicoidal walls. Why such short, non-fibrous cells should have helicoidal wall texture is puzzling. Stone cells show a good range of shapes, including spherical, cylindrical (up to four times longer than wide), and branched.

Pericarp cells in hazel nuts (*Corylus avellana*) have stony walls so thick that they almost obliterate the cytoplasm (Fig. 2.38E, dotted). The drawing shows half a pericarp cell reconstructed from electron micrographs taken by Roland et al. (1989). Many such cells form a nutshell. The laminated cell wall is helicoidal throughout. There are radial pits which traverse the wall layers and communicate between cells via plasmodesmata.

Figure 2.38F shows an epidermal cell from the seed of a quince (*Cydonia oblonga*), reconstructed from electron micrographs by

76 / *The occurrence of fibrous composites*

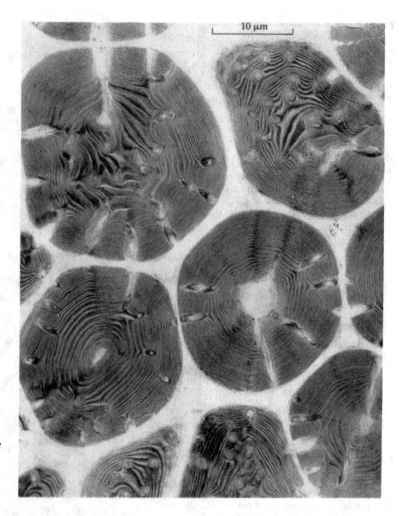

Figure 2.37. Electron micrograph of a section through the stony pericarp (shell) of a nut from a hazel tree (Corylus avellana). *The walls are helicoidal, but at this low magnification only 180° repeats are resolved, not arced patterns. The cell walls are so thick that they occlude the cytoplasm almost entirely. The cells are connected by pits with plasmodesmata. From Roland et al. (1989), by permission of Editions Scientifique, Elsevier, Paris, and of the authors.*

Abeysekera and Willison (1990). The cell is surrounded by walls of neighbouring cells. The huge periplasmic space (black) between cytoplasm and cell wall is partially filled with liquid crystalline mucilage. The liquid crystals contain cellulose microfibrils, some in helicoidal cholesteric array (arced patterns), and some in parallel nematic array (parallel lines). The cytoplasm of the epidermal cell is restricted to a very small volume (dotted). This is thought to form a natural model of a self-assembling plant cell wall.

Figure 2.38G illustrates typical epidermal cells which have helicoidal walls, especially in the thick wall of the outer surface.

Figure 2.38H shows a sketch of root hairs on a root epidermis. The cell walls of such hairs are often helicoidal, especially in aquatic plants (Emons & van Maaren, 1987).

Conifer tracheids have a plywood structured wall like that drawn in Figure 2.38I.

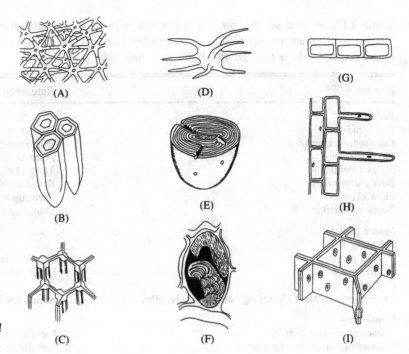

Figure 2.38. Sketches of types of plant cells known to have helicoidal structure in their cell walls. For details, see the text. (A) Parenchyma cells functioning as aerenchyma. (B) Sclerenchyma fibres. (C) Collenchyma cells. (D) Sclerenchyma (sclereid) cell. (E) A pericarp cell (cut through) of a hazel nut shell. (F) Epidermal cell of quince seed with liquid crystals. (G) Typical epidermal cells. (H) Epidermal root hairs. (I) Conifer tracheid.

Carpita (1985) has reasoned that a multilaminated cell wall favours increase in cell size and diversity of cell shapes. Large cells can cope with high tensile forces, which may differ from region to region in a cell of complex shape.

2.5 Pseudo-orthogonal plywoods

2.5a Insect cuticle

We have seen in Chapter 1 that pseudo-orthogonal structure is a combination sandwich of parallel and helicoidal fibril directions. Reference to Table 2.12 shows that this kind of architecture is found in the endocuticle in four separate insect orders (dragonflies, stick insects, bugs, and beetles). This kind of architecture can be visualized in the polarizing light microscope (see Fig. 2.40, leg cuticle of a water bug, *Hydrocyrius colombiae*), but this can resolve only the major layers. A transmission electron microscope is needed to see the helicoidal connecting layers, and an example is shown for *Hydrocyrius* in Figure 1.6. Even a scanning electron microscope cannot resolve the helicoidal layers; only the major fibre directions are seen (e.g. Fig. 2.39, for the wing case of a scarab beetle, *Copris*). The main layer directions for one of the body plates of a giant stick insect (*Heteropteryx*) are seen in the scanning electron micrograph in Figure 2.41. The dominant layers of pseudo-orthogonal cuticles have very precise directions with respect to defined body axes, and they rotate in a systematic way. This is discussed further in Section 5.2k.

Table 2.12. *Selected systems for which electron microscopy indicates pseudo-orthogonal 'plywood' structure, as drawn in Figure 1.17. These systems are like an orthogonal plywood with the major layers connected by a thin sandwich filling of helicoidal layers.*

Species	Reference
Wood	
Pinus radiata	Wardrop & Harada (1965)
Eucalyptus elaeophora	Wardrop & Harada (1965)
Picea jezoensis	Harada (1965)
Pinus virginiana	Tang (1973)
Tilia platyphyllos	Roland (1981), Roland & Mosiniak (1983)
Picea abies	Parameswaran & Liese (1982)
Pinus palustris	Dunning (1968)
Insect cuticle	
Odonata	
Aeshna juncea	Neville (1975a)
Libellula depressa	Jackson (1985)
Phasmida sp.	
Heteropteryx dilatata (forewing, antenna, sternite)	Simmons (1989), Loder (1989)
Hemiptera	
Belostoma malkini (adult)	Neville & Luke (1969b)
Hydrocyrius colombiae (adult)	Neville & Luke (1969b)
Hydrocyrius colombiae (nymph)	Jackson (1985)
Gerridae	Simmons (1989)
Coleoptera	
Scanning electron microscopic survey of families Carabidae, Geotrupidae, Scarabaeidae, Elateridae, Buprestidae, Chrysomelidae, Cerambycidae, Lucanidae, Staphylinidae, Passalidae, Curculionidae, Coccinellidae, Pyrochroidae, Cantharidae, Tenebrionidae	Simmons (1989)

2.5b Wood

Many papers have been published concerning the directions of layers of fibres in woods. There is even a whole book devoted to this (Boyd, 1985). The classical interpretation of the fibre directions in the secondary cell wall of wood tracheid cells dates from the paper by Kerr and Bailey (1934). Tracheids may be modelled by drawing a series of lines (S_1) in parallel and at a defined angle to the long side of a rectangle of clear plastic. This is then rolled up to form a cylinder, converting the lines into parallel helices. The process is repeated for two more sets of helices, oriented at different angles to the axis of the cylinder. The three systems S_1, S_2, and S_3 are secreted in that order, with S_3 being the innermost, and are drawn with the sense of helix (left or right) changing from S_1 to S_2, and then reversing so that S_3 has the same sense as S_1. If made with clear plastic, all three sets of helices may be seen on the model. The angles of S_1, S_2, and S_3 are specific for the

2.5 Pseudo-orthogonal plywoods / 79

Figure 2.39. Pseudo-orthogonal plywood. Scanning electron micrograph of endocuticle from the wing case (elytron) of a scarab beetle (Copris; Scarabaeidae). This is a view of cuticle fractured in the plane of the component layers. The technique shows clearly the major fibre directions but does not reveal the intervening helicoids which connect them. The major directions rotate (apparently) slowly clockwise. By kind permission of M. Jackson (1985).

tracheid walls of a given species of tree (Preston, 1952). S_1 and S_3 are at small angles to the cell length, whereas S_2 is at a large angle to this axis.

In the classical model, there are large angular changes between the layers S_1 and S_2, and also between S_2 and S_3. However, within the past decade it has become clear that there is a thin intervening region of helicoid sandwiched between these layers; as in beetle cuticle, the angular change is gradual, not sudden (see Section 1.3n). Several papers have indicated the existence of small numbers of transition layers between S_1 and S_2 and

80 / *The occurrence of fibrous composites*

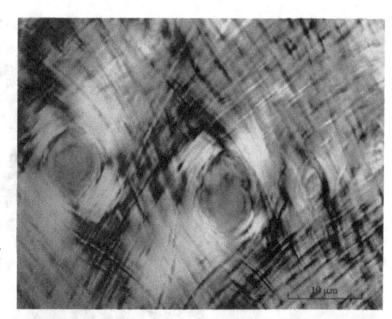

Figure 2.40. Example of pseudo-orthogonal insect cuticle. Surface view in polarized light of two layers of endocuticle cleaved from a leg of a giant water bug (Hydrocyrius colombiae; Hemiptera, Belostomatidae). The major layer directions are seen (but the technique does not reveal the interconnecting helicoidal layers); see Figure 1.17. The rims of the sensory ducts which traverse the cuticle are strengthened by more closely packed and better-oriented fibrils. From Neville (1967c), by permission of Academic Press Limited, London.

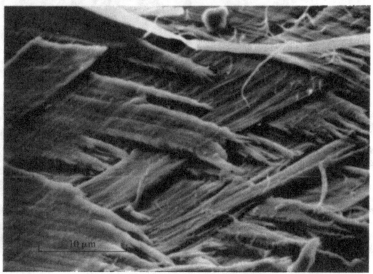

Figure 2.41. Example of pseudo-orthogonal insect cuticle. Scanning electron micrograph of endocuticle from an adult giant Malaysian stick insect (Heteropteryx dilatata; Phasmida). The cuticle of an abdominal sternite has been split open to show the major layer directions. Reproduced by kind permission of P. Loder (1989).

between S_2 and S_3 layers of secondary walls. The initial observations of this kind were made using replicas and layer stripping. Transition layers were recorded in *Pinus radiata* and *Eucalyptus elaeophora* (Wardrop & Harada, 1965), *Picea jezoensis* (Harada, 1965), and latewood tracheids in *Pinus palustris* (Dunning, 1968). Dunning's award-winning paper gives details of methods such as layer stripping and replication. Tang (1973) also deduced the presence of transition layers, using careful birefringence observations on *Pinus virginiana*. This paper is a model example of the power of polarization analysis.

Subsequent transmission electron microscopy of wood sections has revealed that the transitions between S-layers are helicoidal. The realization that this is so was influenced by the growing field of research into animal helicoids and its application to plant cell walls (Section 2.4). Our drawing (Neville & Levy, 1985) to summarize this is given in Figure 1.17, and some examples of pseudo-orthogonal wood cell walls are listed in Table 2.12. These include wood vessels, fibres, and parenchyma in *Tilia platyphyllos* (Roland, 1981; Roland & Mosiniak, 1983) and tracheids in *Picea abies* (Parameswaran & Liese, 1982). A search of the literature revealed further data which we reinterpreted as indicating pseudo-orthogonal structure (Neville & Levy, 1984). Examples include xylem parenchyma in *Populus tremuloides* (Chafe & Chauret, 1974), sieve wall in *Pinus strobus* (Chafe & Doohan, 1972), and xylem parenchyma in *Cryptomeria japonica* (Chafe, 1974). A unifying model for wood cells has been proposed by Roland and Mosiniak (1983). Helicoids in pseudo-orthogonal systems occur in hardwoods (e.g. *Eucalyptus*, *Populus*, and *Tilia*) as well as softwoods. They are found in the secondary walls of vessels and fibres of hardwoods and in the tracheids of softwoods; they also occur in parenchyma, which forms a significant volume of wood.

Confirmation of the helicoidal nature of the transition layers between S-layers has been obtained using the tilting technique (Section 1.3k) by Roland and Mosiniak (1983). The arced patterns seen in lime tree wood (*Tilia platyphyllos*) were altered by tilting with a goniometric stage, as shown in Figure 2.42. In some cases the arcs were reversed.

So the main secondary layers (S-layers) are wound helically around the cell and connected together by helicoidal 'universal plywood'. Wood is a multicylindrical pseudo-orthogonal plywood, with the wall fibres oriented in the sequence S_1, helicoid, S_2, helicoid, S_3. That the fibrous architecture of wood is more complex than previously thought may have implications for those concerned with theoretical estimates of strength and stiffness of timber.

2.6 Anomalous distribution of helicoids

Whilst collecting material for this book it emerged that helicoidal structure may be found in support systems of some taxonomic groups and yet be absent from the same systems in other groups. Such erratic distribution may eventually give clues to the morphogenesis of helicoids.

In fungi, helicoidal structure appears to be restricted to some arbuscular Zygomycetes, which are primitive. They have not so far been found in higher fungal groups such as Basidiomycetes, Ascomycetes, or Phycomycetes (P. Bonfante-Fasolo, personal communication). Also, with the exception of the fungal wall in two lichen species, helicoids are so far unknown in the walls of

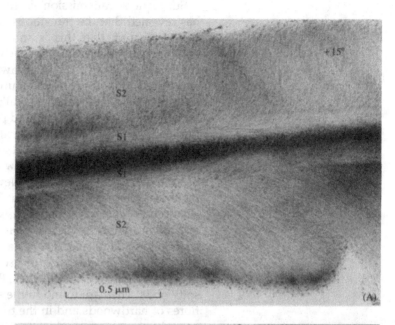

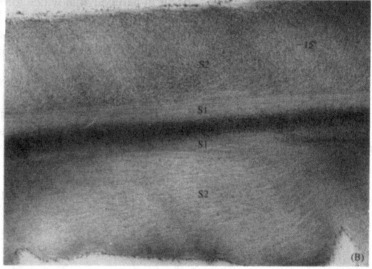

Figure 2.42. *Pseudo-orthogonal plywoods in wood. Electron micrograph of a section through the secondary cell wall of a wood cell from the large-leaved lime tree* (Tilia platyphyllos; *Tiliaceae). Two major layers* (S_1, S_2) *are interconnected by a thin layer of helicoid. The section is tilted on a goniometric stage, changing the appearance of the microfibril directions as predicted by pseudo-orthogonal properties. The same section is seen (A) tilted through +15° and (B) tilted through −15°. From Roland and Mosiniak (1983), by permission of the authors and of the International Association of Wood Anatomists (IAWA), Leiden, Holland.*

fungal hyphae. Perhaps this is because these grow only at their tips and at a rapid pace. *In vitro* data show that cholesteric liquid crystals take time to form (e.g. over a time span of several hours, or even days in mantis oothecal proteins). Tip growth seems incompatible with helicoid formation, which occurs over a large area of cell surface.

The section on female nematode cysts (Section 2.3d) shows that some kinds have round cysts with an inner helicoidal layer, whereas in other quite closely related genera the helicoidal layer is absent. Similarly, helicoidal structure in nematode eggshells seems to be restricted to the order Trichurida.

The chorion in insect eggs is helicoidal in the case of butterflies and moths (order Lepidoptera). No other insect order seems to have this construction, despite the large number of studies on insect eggshells [e.g. see the three volumes by Hinton (1981)]. Helicoidal structure seems an ideal way to strengthen a spherical shell, making it all the more surprising that the method is restricted to a single insect order.

A further surprising feature in arthropods is that whilst all species seem to possess helicoidal structure in the cuticle, only the scarab family of beetles (Scarabaeidae) produce a helicoidal pitch of appropriate size to give interference colours. Other iridescent insects use different methods.

Pseudo-orthogonal architecture is so far known only from the cuticle in four insect orders. It was first recognized in beetles and bugs, giving the impression that it was a feature of evolutionarily advanced groups. It subsequently came as a surprise that pseudo-orthogonal structure is found also in stick insects and in the even more primitive group of dragonflies.

In the molluscs, helicoidal structure in the outermost layer (periostracum) is known so far in only a few species; more information is needed before any suggestions can be made. Another system for which more information is required is the test of sea squirts; it is known that the test of *Halocynthia* is helicoidal, whereas those of some others are vesicular.

In conclusion, helicoidal fibrous composites, as judged by their very wide distribution in both plants and animals, are very important in biology.

3
Properties of natural plywoods

3.1 Chemistry of fibrous composites

In living systems, fibrous composites (defined in Section 1.3d) behave like fibreglass and consist of stiff fibres embedded in a pliable matrix. Their chemistry is dominated by polysaccharides and proteins, both of which are polymers. The chemistry of both the fibrous and the matrix parts of these composites is variable. The following are some examples of the chemical variety, and all possible combinations can occur:

(1) Polysaccharide fibres in a polysaccharide matrix. Examples include plant cell walls (Neville et al., 1976).
(2) Polysaccharide fibres in a protein matrix, such as cuticles of insects, crabs, and other arthropods (Neville, 1975a; Bouligand, 1965), and tests of tunicate sea squirts (Gubb, 1975).
(3) Protein fibres in a polysaccharide matrix, such as eelworm cyst (Shepherd et al., 1972), chicken cornea (Coulombre & Coulombre, 1975), human bone (Giraud-Guille, 1988), and basement lamella (Kefalides et al., 1979). Vertebrate extracellular matrix is not included in detail in this book, as it is already the subject of numerous works (Hay, 1981; Trelstad, 1984; Bard, 1990).
(4) Protein fibres in a protein matrix, such as fish eggshells (Grierson & Neville, 1981), butterfly and moth eggshells (Smith et al., 1971), and praying mantis eggcase (Neville & Luke, 1971b).

Examples of polymers which form the fibrous component of composites include the polysaccharides cellulose and chitin and the protein collagen. Matrix components include the polysaccharides hyaluronic acid and hemicellulose and the protein rubber resilin. There is a tendency to consider hyaluronic acid as a molecule characteristic of vertebrate materials. It is, however, also found associated with collagen type IV in the basal lamina of insects and in the material surrounding a regenerating leg in a stick insect, *Carausius* (Browaeys-Poly, 1991). Similarly cellulose, while typical of plant cell walls, is also found in the test of sea squirts, where it was formerly known as tunicin.

3.1a Polysaccharide chemistry

Some polysaccharides are simple in construction, polymerized from only one type of residue such as polyglucose (cellulose) and poly-N-acetylglucosamine (chitin). The residues are covalently bonded by strong $\beta(1 \to 4)$ linkages.

Slightly more complex are those unbranched extracellular polysaccharides which contain more than just one kind of sugar residue. These are termed mucopolysaccharides or glycosaminoglycans. The different residues normally alternate along the chain and usually consist of an amino-sugar (often acetylated, e.g. N-acetylglucosamine) and one of the forms of uronic acid. The constituents of the major ones are as follows; hyaluronic acid (N-acetylglucosamine alternates with glucuronic acid), chondroitin (N-acetylgalactosamine alternates with glucuronic acid), dermatan (N-acetylgalactosamine, idiuronic acid, and glucuronic acid), and keratan (N-acetylglucosamine alternates with galactose). These polysaccharides are often combined with proteins to form glycoproteins (proteoglycans).

Some polysaccharides are much more complicated, with short sidechains branching off the backbone chain (e.g. plant hemicelluloses, Fig. 5.2). The shape of such molecules may be likened to pipe-cleaners. They may have backbones either of glucose (forming a glucan) or of xylose (forming a xylan). The sidechain residues have linkages which are more flexible than the rigid $\beta(1 \to 4)$ backbone links. Sugars involved in hemicellulose sidechains include arabinose, fucose, galactose, and xylose. Because of the sidebranching, hemicelluloses are even more diverse than proteins (which never branch), despite the fact that they are built from fewer kinds of monomers (twelve kinds of sugars versus twenty kinds of amino acids).

3.1b Protein chemistry

Knowledge of the structural proteins of fibrous composites is in many cases at the stage of isolating them and rearing antibodies. These, when labelled with gold, then reveal the specific locations of the proteins in electron micrographs of thin sections. Information of this kind is now available for many of the different proteins of insect cuticles, which number in excess of 100 in each particular sample. The insects being analysed include the locust *Locusta migratoria* (Andersen, Hojrup, & Roepstorff, 1986; Andersen & Hojrup, 1987), the silk moth *Hyalophora cecropia* (Cox & Willis, 1985), the tobacco hawk moth *Manduca sexta* (Kiely & Riddiford, 1985), the mealworm beetle *Tenebrio molitor* (Lemoine et al., 1990), the sheep blowfly *Lucilia cuprina* (Skelly & Howells, 1988), and the fruit fly *Drosophila melanogaster* (Chihara, Silvert, & Fristrom, 1982; Wolfgang et al., 1986). The blend of proteins varies with metamorphosis from larva to pupa to adult, with time of deposition (premoult or postmoult), and with different parts of the exoskeleton.

Most of the effort by protein chemists has naturally enough centred on the analysis of dynamic molecules like respiratory proteins and enzymes. Not many sequences are known for extracellular structural proteins. There are several levels of analysis for proteins. Of these, amino acid composition tells us very little. Even an amino acid sequence may not reveal enough about protein shape. What is required is knowledge of secondary structure and conformation. This information is, however, known for collagen (Walton, 1975), and it reveals a predominant regular repeat of three amino acids: glycine, proline, and hydroxyproline. The last two forbid the formation of an α-helical secondary structure and cause instead the formation of a triple helix, several of which then organize into a rope-like microfibril.

The composites for which most is known are the chorion proteins of moth eggshells. We produced the incentive to study this system by showing that it self-assembled into a helicoid (Smith et al., 1971). Another advantage is that there is no chitin present – only proteins. A massive research programme at Harvard headed by Professor F. C. Kafatos has now sequenced the almost two hundred proteins involved. Work is proceeding (Hamodrakas, Bosshard, & Carlson, 1988) to establish the secondary structure and to relate this to the formation of the helicoidal architecture (Section 5.1b).

3.1c Interfaces between fibres and matrices

This section should be read in conjunction with Section 3.2 on mechanical functions. Strong chemical linkage between the fibrous and matrix components of composites is vital to their mechanical performance. In manufactured fibreglass, the bond between the glass fibres and epoxy resin is hydrophilic, making it susceptible to attack by water. This is counteracted by a coupling agent, with small molecules such as *silanes* attached at one end to the glass, and at the other end to the resin (Sterman & Marsden, 1968; Wong, 1968). What do we know about the interfaces in biological composites?

It has been almost half a century since Fraenkel and Rudall (1947) noticed the compatibility between the repeat distances along a chitin chain and along a protein chain in the β-extended (pleated) conformation. It turns out that the repeat of four aminosugars in chitin (2.056 nm) is equal to the repeat of six amino acids in β-extended protein. I reproduced their diagram of this (Neville, 1967c), but without realizing its full significance, despite the prior publication of a composite model for cuticle (Jensen & Weis-Fogh, 1962). It was only with the observation of chitin

88 / *Properties of natural plywoods*

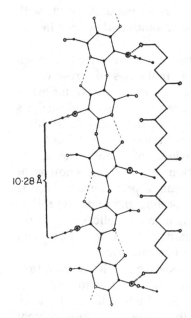

Figure 3.1. Diagram of the probable bonding between chitin and protein in insect cuticle. The compatibility of chitin (on the left) and β-pleated region of a cuticle protein (on the right) was first noted by Fraenkel and Rudall (1947) on the basis of X-ray diffraction evidence. Every fourth amino-sugar residue (acetylglucosamine) coincides with every sixth amino acid residue. The points of coincidence are connected by hydrogen bonds (dotted) between N–H (circle with a dot) on the chitin, and C=O (open circle) on the protein.

crystallites (Rudall, 1965), and their positioning in different types of chitin architecture (Neville & Luke, 1969b), that the importance of Fraenkel and Rudall's observation emerged. A more detailed diagram of the chitin/protein interface is given in Figure 3.1.

The method used to show the coincidence of repeat distances (Fraenkel & Rudall, 1947; Rudall, 1963) was fibre X-ray diffraction of insect cuticle samples in which the chitin and some of the protein were oriented in parallel (e.g. along tendons and apodemes). The protein was then chemically removed, producing a corresponding decrease in density of the appropriate X-ray reflections on the fibre axis.

It is not envisaged that all of the proteins in insect cuticle are β-extended in conformation (as they are in, for instance, silk fibroin). Proteins in the matrix could bond to the appropriate face of a chitin crystallite (see Fig. 1.3A) even if only part of each protein was in the β-pleated sheet conformation.

Blackwell and Weih (1980) constructed a possible model of the interfacing between chitin and protein in an insect cuticle (Fig. 3.2). This was based on X-ray diffraction studies of the parallel chitin chains in the wood-drilling ovipositor of an ichneumon fly (*Megarhyssa*). In the model, a six-fold helix of protein units is arranged as a sheath around a central chitin crystallite core, whose diameter is 2.8 nm, as measured by Neville et al. (1976). Protein units of this size would have a molecular weight of about 27,000. This was the first attempt to model a fibre/matrix interaction in a biological fibrous composite.

Another example from insect cuticle is more puzzling. It concerns the protein rubber resilin, whose 97% elastic efficiency indicates that its chains are oriented randomly (Andersen & Weis-Fogh, 1964). Resilin sometimes occurs in the pure form (rubber ligaments), storing kinetic energy by elastic deformation during insect flight and flea jumping. But it also occurs reinforced with chitin crystallites (Fig. 3.3 and see Fig. 1.5). In some locust ligaments the chitin is helicoidally arranged, and resilin is the only other component (Neville, 1975a). Such a composite could not work unless the chitin was firmly bonded to the resilin matrix. The number of hydrogen bonds in a thermodynamic rubber such as resilin is reduced to a minimum (Andersen & Weis-Fogh, 1964); otherwise the elastic efficiency would be reduced. Is part of the resilin chain in a β-pleated sheet conformation, and if so, is that part bonded to chitin as in Figure 3.1?

It has been proposed that hemicelluloses in plant cell walls self-assemble into a helicoidal matrix (Neville, 1985a), which hydrogen-bonds to cellulose crystallites. In so doing, the hemicellulose could bring about the orientation of the cellulose. This topic is explored in Section 5.1a. It is known from *in vitro* studies that cellulose and hemicellulose do in fact hydrogen-bond together (Bauer et al., 1973), with the chains in parallel. This is enabled by both polymers being β(1 → 4)-linked along their backbones, so that their residues lie in register.

Collagen is found in association with glycosaminoglycans, but the interaction occurs not at the level of the collagen molecular chain but rather at the collagen fibril level. In some systems (e.g. basement lamella, cornea) the glycosaminoglycan matrix may possibly position the collagen fibrils in helicoidal array; there is ample scope for further work here both *in vivo* and *in vitro*. Extracellular matrix, with the same types of molecules, is found also in the basement membrane of insects and surrounding regenerating limbs in stick insects (Browaeys-Poly, 1991).

We may detour briefly inside the cell to admire the effective way in which the β-keratin of feathers forms a composite with a built-in covalent interface between fibrils and matrix. The central part of a β-keratin molecule folds into a β-pleated sheet, while the outer parts form an irregular shape (Fig. 3.5). The β-pleated regions of neighbouring molecules hydrogen-bond together to form a microfibril; the less regular regions form the matrix. So the problem of interfacing fibrillar and matrix components does not arise. They are joined by a covalent peptide bond during protein synthesis on the ribosomes in the normal way. It seems unlikely that feathers could achieve what is demanded of them during flight without this molecular adaptation.

A synthetic example of this principle is found in three-block copolymers such as SBS (polystyrene, polybutadiene, polystyrene), in which the outer arms of the chain crystallize with neighbouring arms to form 15-nm-diameter polystyrene crystallites set in a matrix of butadiene formed by the central regions of the copolymer (Keller et al., 1971). The stiff crystallites are distributed throughout the rubbery matrix in a hexagonal array, with a centre-to-centre separation of 30 nm.

Some extracellular composites have interfaces between fibre and matrix which are based upon the same principle as feather keratin and SBS. The secondary conformation of moth eggshell chorion proteins (Hamodrakas et al., 1988), and also of fish eggshell proteins (Hamodrakas et al., 1987), is like that of feather keratin. Here again different parts of the same protein chain contribute (along with neighbouring chains) to form both fibres and matrix (Fig. 3.5).

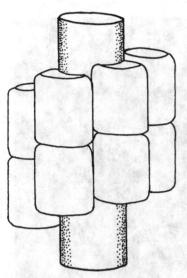

Figure 3.2. *Diagram of a model of a possible interface between chitin and protein in insect cuticle. Protein units are arranged in a six-fold helix around a central chitin crystallite core (diameter 2.8 nm). In this side view, some of the protein units are obscured. This was the first attempt to model a fibre–matrix interaction in a biological fibrous composite. Redrawn from Blackwell and Weih (1980).*

3.2 Mechanical functions: fibre orientation strategies

Three papers were influential in creating interest in composite materials in animals. These were Currey (1962), on bone, and Jensen and Weis-Fogh (1962) and Weis-Fogh (1965), on insect cuticle.

90 / *Properties of natural plywoods*

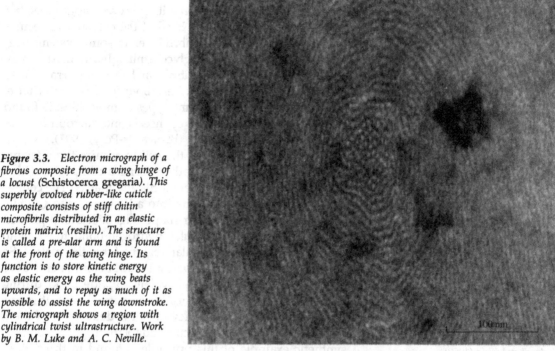

Figure 3.3. Electron micrograph of a fibrous composite from a wing hinge of a locust (Schistocerca gregaria). This superbly evolved rubber-like cuticle composite consists of stiff chitin microfibrils distributed in an elastic protein matrix (resilin). The structure is called a pre-alar arm and is found at the front of the wing hinge. Its function is to store kinetic energy as elastic energy as the wing beats upwards, and to repay as much of it as possible to assist the wing downstroke. The micrograph shows a region with cylindrical twist ultrastructure. Work by B. M. Luke and A. C. Neville.

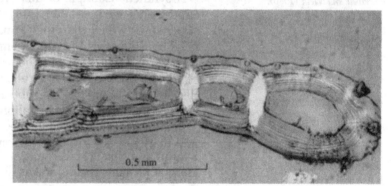

Figure 3.4. Transverse section through a hind tibia of a giant water bug (Hydrocyrius colombiae; Hemiptera, Belostomatidae), photographed between crossed polaroids. The tibia is flat in shape for use as a swimming oar and is strengthened by bars (white in the photograph) of cuticle, which dowel the two sides of the structure together. Similar pegs dowel together the upper and lower cuticles of beetle elytra. The chitin microfibril orientation in the endocuticle of the oar is pseudo-orthogonal except in the pegs, where it is parallel to the length of the pegs. From Neville (1967c), by permission of Academic Press Limited, London.

3.2a Fibrous composites

The principle of fibrous composites, of which fibreglass is an example, is based upon two conflicting properties of fibrous materials. Glass fibres illustrate this well because they combine high stiffness (very low extension even when stretched by high tensile force) with brittleness. They are brittle because of the scratches which inevitably occur on the surface of glass. Forces concentrate upon scratches so as to open up cracks (known to

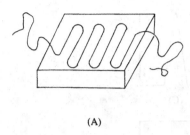

(A)

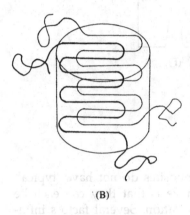

(B)

Figure 3.5. Highly diagrammatical sketch of the molecules forming the fibrous composite in moth eggshell. (A) Each molecule has three regions. The central part is a β-pleated sheet; the two outer arms have a random conformation. Redrawn from Hamodrakas (1984). The β-sheets are probably not planar as shown here, but twisted as in Figure 5.6. (B) The central domains of several molecules crystallize together to form a stiff microfibril, whereas the outer arms form a more rubbery matrix. This principle occurs in β-keratin of feathers, and probably also in the extracellular proteins of moth eggshells. The β-sheets of the central domains probably are not planar, as drawn here for simplicity, but twisted as shown in Figure 5.6. In this system, the problem of bonding fibre to matrix never arises, because the arms are joined to the central region by peptide bonds formed in the usual way on the ribosomes during synthesis of the protein chains.

engineers as Griffith cracks). The tip of a crack is extemely sharp – it passes, after all, between the atoms of a substance which is breaking. A tensile force focuses upon the small surface area of a crack's leading edge, giving a very high stress. (Stress is force per unit area; hence the smaller the surface area of a crack, the higher the stress acting on its edges.) Vincent (1982), in his clearly written book on biological materials, gives the example of a glass-cutter scoring a pane of window glass with a diamond. The glass may subsequently be broken along this defect, because stress concentrates on its edges. In summary, the high stiffness of glass is potentially advantageous, but its brittleness is detrimental.

The construction of fibrous composite materials overcomes this problem by combining fibres of high stiffness with a matrix of very low stiffness. In fibreglass this is done by setting the glass fibres in a resin matrix. A potentially damaging crack running through a composite then uses up some of its energy in stretching the deformable matrix. The crack eventually fizzles out, and the composite survives. Paradoxically, the apparent dilution of a strong material (glass) with a weaker one (resin) creates a composite material which is stronger than a pure sample of either.

In engineering terms, 'stiffness' is called Young's modulus (E). This is equal to stress (σ)/strain (ε). Stress is force per unit cross-sectional area. The same force acting upon two samples of the same material which have different areas of cross-section will produce a higher stress in the sample with lower cross-sectional area. Steady stress is measured in force in Newtons (mass in kilograms × acceleration due to gravity in metres per second per second) per area of cross-section (in square metres). The units of stress are Newtons/metre^2 (N/m^2). Strain is the resulting percentage elongation due to a stress, and thus has no units. Hence Young's modulus (E) = σ/ε, in N/m^2. Materials with low E are very extensible; those with high E are stiff and inextensible.

An ideal combination in a biological system is that of the elastic protein resilin (not to be confused with the resin of fibreglass) with stiff crystallites of chitin. The value of E for resilin is very low (2×10^6 N/m^2), and that for chitin very high (9×10^{10} N/m^2). Sometimes resilin occurs pure, but often it forms a matrix for chitin crystallites (Figs. 1.5 and 3.3). Both pure and composite versions occur in the insect flight system and flea jumping system, where they repetitively store and release energy with great efficiency. I have flown locusts to fuel exhaustion in a wind tunnel for twenty hours. During that time they have executed over one million wingbeats, with no noticeable deterioration in performance. Fleas will carry on jumping if confined in small numbers in a box. The number of jumps may be counted by a microphone and counter. As many as ten thousand total jumps have been recorded by ten fleas in an hour (Rothschild et al., 1973, 1975).

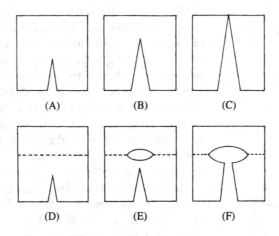

Figure 3.6. How laminates and fibrous composites work. (A–C) A sequence of diagrams of the passage of a crack through a sample of pure material (i.e. no laminated interfaces). The damaging sharp crack runs unrestricted through the whole material. (D–F) By contrast, this sequence shows a crack running through a material which has a weak laminated interface (dashed line). A secondary hole opens up at the weak interface and blunts the crack tip by absorbing some of the strain energy which drives the primary crack. The composite material survives cracks which would cause total failure in a non-laminated counterpart. So fibre–matrix interfaces are important in the functioning of fibrous composites, and multiple interfaces (as in plies and helicoids) make a laminate very strong. The diagrams illustrate the Cook–Gordon model.

3.2b Variable factors in composites

As Gordon (1988) has noted, composites do not have 'typical' properties. The advantage they have is that they can each be designed for a specific task and location. Several factors influence the mechanical performance of fibrous composites. The smaller the diameter of the fibrous component, the more will the energy of potentially damaging cracks become dispersed throughout the matrix. Long fibres are an advantage wherever they are used in parallel (e.g. in animal tendons or plant stem fibres). Composite properties also vary with the volume fraction of fibres to matrix. Also, the neater the distribution of the fibrous component to form a regular lattice (e.g. hexagonal array of chitin crystallites in some insect cuticles), the less mechanical defects will occur in the system.

It is important to have strong chemical bonding between the fibrous component and the matrix (see Section 3.1c). Provision of a protein sheath around each chitin crystallite in insect exoskeletons will protect its surface from defects. Likewise, a hemicellulose sheath protects cellulose crystallites in plant cell walls. But there is another principle at stake: that of *lamination*, and this is prevalent in biological materials (Chapter 2). The way laminates work is described by the Cook–Gordon model (Fig. 3.6). They depend upon the insertion every so often of laminar weak interfaces. When a crack runs into a weak interface, oriented perpendicularly to its direction of propagation, a hole opens up. This increases the surface area of the crack and distributes its energy in a less concentrated manner; the secondary crack absorbs some of the strain energy driving the primary crack, and so blunts its edge. The occurrence of multiple weak interfaces makes a laminated material stronger. The mechanics of materials is full of paradoxes.

The ways in which laminated composite materials may fail mechanically are as follows: (1) by delamination (the layers split apart), (2) by buckling (this may be counteracted by cross-graining of plies), (3) by failure of the matrix, and (4) by breakage of the fibres.

The most important variable in composite materials is strategic orientation of the fibrous component. This can overcome the problem of the difference in stiffness *along* versus *across* a fibre or crystallite. Fibres are strong along their length, but weak across the width. A tensile force acting perpendicularly to the fibres in a parallel-fibre composite will be acting on the weak matrix between the fibres. If, however, the fibre directions change from layer to layer, they divert the path of a crack. More crack energy is absorbed by the increasing surface area of the crack, and the fracture strength is improved. The remainder of this section is about the strategies used by plants and animals to make their fibrous composites as effective as possible.

3.2c Manufactured plywoods

Industrial plywoods have the grains of each two adjacent plies oriented mutually at 90° (Fig. 3.7). This orthogonal cross-lamination prevents splitting. Plywood is best employed in situations where shear forces are acting (e.g. tennis rackets, golf clubs, and hockey sticks). When, however, forces act from a constant direction (e.g. bookshelves), natural wood is best. Plywood is an ancient invention; it is even found in the casket of the Egyptian king Tut-Ankh-Amon. For a comprehensive account of plywood and its manufacture, see Sellers (1985).

Synthetic composite plywoods are now being manufactured by British Aerospace (Fig. 3.8). One of these employs 60% volume fraction of carbon fibres set in a thermoplastic matrix. The carbon fibres are wound in several different directions. For a review of the design of modern industrial composites, see Chou, McCullough, and Pipes (1986).

3.2d Theoretical stiffness of composites (parallel, orthogonal, and helicoidal)

The efficiency (η) of reinforcement contributed by fibres in a composite depends upon their direction with respect to applied stress. In an orthogonal plywood,

$$\text{stiffness } E\eta = \sum an \cos^4 \phi$$

where E is Young's modulus of the whole system, η is the efficiency factor in a defined direction, an is proportion of fibres,

94 / *Properties of natural plywoods*

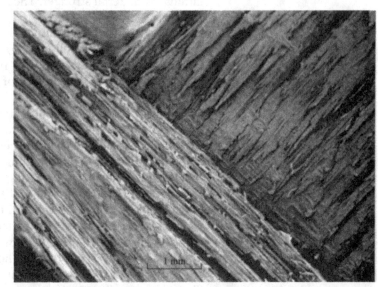

Figure 3.7. Scanning electron micrograph of a manufactured plywood. The sample is broken to show two plies of wood, glued together at 90°. Micrograph courtesy of S. Martin.

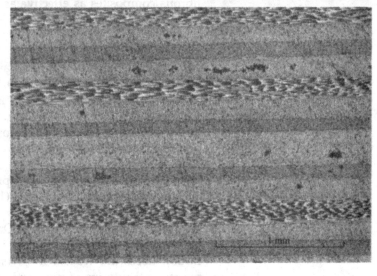

Figure 3.8. Light micrograph of a sectioned surface of a synthetic fibrous composite with multi-ply structure. This is a thermoplastic reinforced with a 60% volume fraction of carbon fibres. This fibrous composite has carbon fibres (7 μm in diameter) with high stiffness (E) set in a thermoplastic matrix of low stiffness. Fibre direction is varied during manufacture to make a plywood texture. Reproduced by kind permission of the Sowerby Research Centre, British Aerospace plc, Bristol, U.K.

microfibrils, or crystallites in the line of pull, and ϕ is the angle of each specific component layer of fibres to the direction of pull. (Krenchel, 1964).

The formula is best illustrated by examples, beginning with the simplest. *Example 1* is a parallel bundle of fibres pulled axially in tension. Here an = 1, and $\phi = 0°$, so $\cos \phi = 1$. Hence the efficiency of fibres $\eta = 1$. This will occur in the chitin crystallites of insect tendon or the collagen fibrils of vertebrate tendons.

In *example 2* the same parallel bundle is pulled perpendicularly to its alignment. Here an = 0, and $\phi = 90°$, so $\cos \phi = 0$. Hence efficiency $\eta = 0$.

In e*xample* 3 an orthogonal ply made from layers of equal thickness is pulled in parallel to the grain of one of the ply directions. Here an = 0.5, $\phi = 0°$, so $\cos \phi = 1$. Efficiency $\eta = 0.5 \cos^4\phi = 0.5$. This would occur in the cellulose fibres of *Oocystis* cell wall.

In *example* 4 the same orthogonal ply is pulled at 45° to both grain directions. Each set of fibres bears $0.5 \cos^4 45° = 0.5 \times 0.25$. The total efficiency thus reduces to 0.25.

For a helicoid (which is a universal plywood), I introduced a term (Neville, 1985a) to adapt Krenchel's formula: an = χ/π rad, where χ is the angle between neighbouring planes in the helicoid. The formula for a helicoid becomes

$$E\eta = \sum \chi \cos^4\phi / \pi \text{ rad.}$$

Testing for $\chi = 15°$, χ/π rad = 15°/180° = 1/12. It is necessary to calculate $\cos^4\phi$ for each angle of ϕ. This gives an efficiency η for helicoids of 0.375. This is applicable to all helicoids except where distortion has occurred (e.g. by growth of a long plant cell wall, or by forces occurring in a moulting insect). The value is independent of helicoid angle χ, of helicoidal pitch, and of direction of pull (provided that it is in the plane of the component plies). So, in a helicoid, although the stiffness is only 37.5% of that of a parallel tendon made of the same fibrous composite, it nevertheless conveys the benefit of being equally stiff when subjected to stress in any direction perpendicular to the helicoidal axis of rotation.

It is predicted from this calculation that an insect cuticle which is helicoidal throughout its thickness (e.g. the caterpillar of *Manduca sexta*) should swell equally in all directions in the plane of the helicoidal plies. This has been demonstrated with strips cut either circumferentially or axially to the larval body length (Wolfgang & Riddiford, 1987).

It is possible to trace the reduction in Young's modulus for the various hierarchical structures in wood (Jeronimidis, 1980). This proceeds from cellulose crystallite to cellulose fibre, to cell wall, to wood. The contribution of helicoids to these estimates can now be carried out, as can similar calculations for the helicoidal components in bone osteons.

3.2e Functions of helicoids

In vertebrates, with the exception of mammals, it seems as if at least part of the thickness of the cornea consists of helicoidal layers of collagen fibres (Coulombre, 1965). Examples are now known from fish (carp and goldfish), amphibians (frog), reptiles (turtle), and birds (domestic fowl, quail, pheasant, and duck), as cited in a preliminary note by Trelstad (1982). It is likely that the helicoids function during development to make and maintain (aided by intraocular pressure) a spherical shape to the cornea,

Figure 3.9. *Locations of helicoids in a plant stem. A diagram to show the locations of tissues with helicoidal cell walls in a transerse section of a dicotyledonous stem (Labiatae). The helicoidal walls are placed strategically for strengthening the stem, with epidermis around the outside (cells indicated), collenchyma (circles), and sclerenchyma (black). These tissues are arranged according to the tubular construction principle – around the outside of the stem.*

thus reducing spherical aberration. This would be expected of a helicoid, which has equal stiffness in all directions.

The proteins which encase the eggs of praying mantids are exuded around them in the form of a cholesteric liquid crystal. This is then spun into a foam by the cercal appendages. Gubb (1976) has shown that when the extracted proteins are added to soap bubbles (SDS), they prolong their life before bursting. Perhaps they also stabilize the mantis foam? It is likely that a cholesteric film would stabilize if its thickness consisted of an integral number of half-turns of helicoid.

When a beam is bent, the greatest stresses are concentrated at its surface. The beam can thus be hollowed out to save weight, while hardly affecting strength. This tubular principle is used in scaffolding and tubular furniture. It therefore makes sense that helicoidal structures are commonly found in peripheral body locations, in both plants and animals. Thus helicoids in higher plants (Fig. 3.9) are commonly found in the epidermis, which covers the outsides of leaves and stems, in bundles of collenchyma, and in the sclerenchyma associated with the vascular bundles. Helicoids would not be expected in the centres of stems or roots, which carry virtually no stresses at all. (Aerenchyma is an exception – see the next paragraph.) In arthropods, the mechanical stress-bearing layers are covered on the outside by a very thin epicuticle, whose functions concern other problems such as water regulation and pheromone recognition. Within this, the exocuticle is almost exclusively helicoidal. Helicoids are well suited to resist shearing forces. Vincent (1981) has shown how helicoidal cuticle stabilizes the outermost layer of the extensible abdominal intersegmental cuticle of adult female locusts. This cuticle is capable of amazing extension to fifteen times its original length, with almost complete recovery. Such cuticle, found only in females, is superbly adapted to permit extension of the abdomen for the purpose of laying eggs as deeply in the sand as possible. The helicoidal component acts as a protection against the shearing forces which occur during extension and recovery.

An exception to the principle that helicoids are located peripherally is in the cell walls of rush pith (Fig. 3.10), first illustrated by scanning electron micrographs by Roland (1981). Pith (also known as aerenchyma) is centrally located in the leafless stems, where only small stresses operate. This paradox is perhaps best explained by a combination of factors. The star-shaped (stellate) cells which join to make pith produce an open network which encloses a complex of air spaces. This results in maximum strength combined with minimum weight. The stems are light and buoyant and can float if their habitat is flooded. The air spaces allow gaseous oxygen to diffuse down to the roots, which are located in wet and stagnant soil. The rushes thus enrich the soil around them with oxygen.

Rush aerenchyma seems to be an example of overkill in mechanical design. It illustrates four mechanical principles, and thus forms a paradigm model for teaching purposes: (1) The stellate cells have tubular extensions. (2) The cell walls are made as a fibrous composite. (3) The cellulose microfibrils are helicoidally arranged (Roland, 1981). (4) The whole pith system forms a geodetic framework. (Geodetics are discussed in Section 3.2g.) The odd thing is that while the pith of *Juncus effusus* is continuous, that of other species is intermittently broken. Pith with all of the foregoing attributes is also found in ancient papyrus (Mosiniak, 1986; Mosiniak & Roland, 1986), which was made from *Cyperus papyrus* in biblical times. British Aerospace manufactures a non-structural space-filling network which closely resembles rush aerenchyma. With their kind permission, I have grouped together scanning electron micrographs of these two systems for comparison (Figs. 3.10 and 3.11).

In plants, helicoidal cell walls will resist turgor pressure equally in all directions, like the cornea of the lower-vertebrate eye resists the pressure of fluid from within the eye.

3.2f Function of pseudo-orthogonal systems

These are complex systems (see Sections 1.3n and 2.5), found in some insect cuticles and more recently recognized in wood. It is likely that the thin regions of helicoids which interconnect the major layer directions (Fig. 1.17) would extend the surface area of cracks running through them – and hence increase fracture strength. The orientations of the main layers in beetle cuticle are accurate to a surprising ±3° (Zelazny & Neville, 1972). They are probably designed to resist forces in specific directions, though a full-scale analysis has yet to be attempted. Likewise, in wood there is great precision of orientation of the major layers S_1, S_2, S_3 (Preston, 1974). Vincent (1982) argues that the angle which the S_1 crystallites make to the long axis of a wood cell is a compromise. If the angle is large, the cell can buckle, making it able to absorb more stress energy, but at the expense of lower axial stiffness. The opposites apply for small angles. We need someone to calculate the effects of the helicoidal transitions between the S-layers in timber. This is complicated by the three S-layers making different angles with the cell axis.

3.2g Geodetic fibres

Fibres which are arranged in a geodetic manner take the shortest distance between two points (A and B) on a curved surface. If a tensile stress acts between points A and B, the geodetic fibre will take the strain *immediately*. This would not happen if the fibre did *not* run the shortest distance from A to B (i.e. if it was

98 / *Properties of natural plywoods*

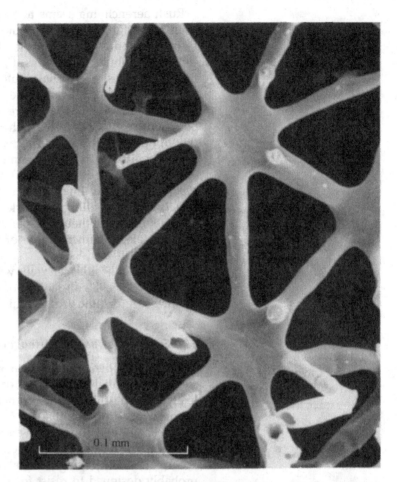

Figure 3.10. Scanning electron micrograph of the pith of a rush stem (Juncus effusus). The stellate cells of the aerenchyma form a network which seems almost too strong for its job – an example of mechanical overkill? It shows four mechanical adaptations: (1) Its cell walls are built as a fibrous composite, with cellulose microfibrils set in a matrix of other polysaccharides. (2) Its walls have helicoidal structure (Roland, 1981). (3) The cell wall extensions are tubular, combining strength with lightness. (4) The strain is taken up by the geodetic arrangement of tubular cell walls. The reason for all this seems to be to create a large volume of air so that oxygen can diffuse readily down to the roots (which grow in stagnant earth), whilst retaining mechanical stability in the stems. Work by S. Martin and A. C. Neville. Compare with Figure 3.11.

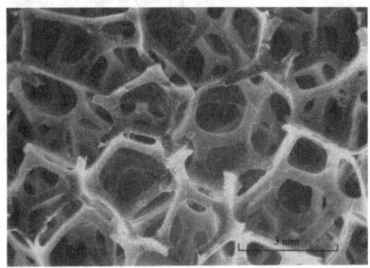

Figure 3.11. Scanning electron micrograph of a non-structural synthetic space-filling network. By kind permission of the Sowerby Research Centre, British Aerospace plc, Bristol, U.K. Compare with rush pith in Figure 3.10.

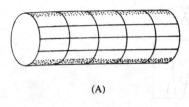

(A)

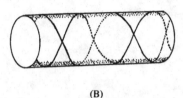

(B)

Figure 3.12. Diagrams of pressurized cylinders with geodetic fibres. (A) Orthogonally oriented layers of fibres lying parallel and perpendicular to a cylinder axis. This kinks if bent and does not allow changes in cylinder length or diameter. (B) A cylinder with layers of fibres wound as alternate left and right helices. This cylinder will bend smoothly without kinking. Changes in length and diameter are also possible. Redrawn from Wainwright (1988).

actually curved). A tensile stress acting on such a curved fibre would merely take up some of the slack – straightening out rather than being stretched.

We may extend this concept to layers of fibrous plies which are arranged geodetically on the surface of a cylinder (Fig. 3.12). In Figure 3.12A, one layer of parallel fibres is drawn parallel to the cylinder axis, the other perpendicular to it. In Figure 3.12B, one layer of fibres is drawn wrapped around the cylinder as a left-handed helix, the other as a right-handed helix. If such a cylinder was cut open down its long axis and spread out, the helices would appear as straight lines. (Had they not been geodetic the fibres would have been curved.) Wainwright (1988) distinguishes between the mechanical behaviours of two cylindrical animals whose body walls have fibres arranged as in Figure 3.12. In the case of Figure 3.12A, the worm would kink if it bent, because the geodetic fibres would allow it neither to extend in length nor to thicken in diameter. But in the other worm (Fig. 3.12B), changes in both length and diameter are possible, and the worm can bend smoothly without kinking.

This principle is illustrated by the examples in Figure 3.13. In the nematode roundworm *Ascaris*, layers of parallel and geodetic collagen fibres are arranged as left and right alternating helices (Fig. 3.13A); if unrolled as in Figure 3.13B, the collagen fibres are seen to be straight. Harris and Crofton (1957) showed the significance of the angle made by the geodetic collagen fibres with the long axis of the body; in *Ascaris* this angle is 76°. Had the angle been small, the fibres would have had an effect similar to that of longitudinal fibres, preventing the worm from lengthening. But because the angle is large, the worm is prevented from shortening. This enables the longitudinal muscles on the dorsal and ventral sides of the body to work against each other, bending the worm – there are no circular muscles. The critical angle of fibres to long axis is 55°. Above this value, worms cannot shorten; below it they cannot lengthen. It all depends upon the species.

The geodetic principle also applies to helicoidal systems such as insect cuticle. In Figure 3.13C, a cylinder of insect cuticle is drawn with just a few directions of helicoidal chitin crystallites indicated. If unrolled, the geodetic crystallites should be straight, as in Figure 3.13D. Occasionally, cases are encountered where the crystallites are actually curved in the plane of the cuticle, as in Figure 3.13E, for example, in some stick insects (Dennell, 1976). It would be of interest to study the occurrence and relative abundance of geodetic and non-geodetic crystallites in arthropod cuticle.

At a specified location in a helicoidal system, stiffness and fracture strength have equal values in all directions normal to the axis of helicoidal rotation; they effectively have radial symmetry. But in many other cases, sheets of fibres are oriented

Figure 3.13. Diagrams of geodetic mechanical strategies in fibrous cuticles. (A) The cuticle of the nematode Ascaris contains collagen fibres oriented at 76° to the worm's long axis (Harris & Crofton, 1957). Because the worm is cylindrical, these appear as alternating right- and left-handed helices. (B) When drawn in unrolled view, it is seen that the collagen fibres are in reality straight. They are therefore geodetic and take the strain immediately the worm bends. (C) An imaginary cylinder of arthropod cuticle is drawn to show a sample of four superimposed chitin fibre directions. These, together with many more, combine to make up a helicoid. That the fibres of the helicoid are geodetic (shortest distance between two points on a curved surface) is shown when the cylinder is unrolled. (D) If the fibres were really curved, they would still appear so in the unrolled view (part E). Therefore, when a force acts on a cuticular helicoid, some of the fibres are immediately able to take up the strain, no matter the direction of the force. Had they been really curved, this would not be possible. The principle in parts C, D, and E has yet to be demonstrated on arthropod material.

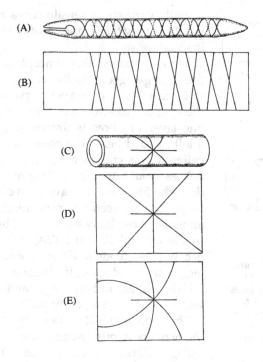

alternately in left and right directions, balancing out to function as a bilaterally symmetrical system. Examples include the collagen tunic which encloses the mantle muscles of a squid (Fig. 3.14A) and the collagen of the basement membrane of a tadpole epidermis (Fig. 3.14B). For further examples of such systems, see Alexander (1987, 1988).

3.2h Spherical shells

One type of situation where helicoids are found commonly is in the walls of the resting stages of plants and animals which have spherical shells. Examples are known from a very wide range of phyla (Table 2.7), as indicated by arced patterning in electron micrographs. In particular it seems that helicoids may be associated with dormant stages engaged in dispersal when their temporary habitats – such as freshwater or brackish water, dust or soil – dry out. Some of the examples may enter prolonged desiccation (e.g. rotifer and nematode cysts). In others, such as the protistan *Actinophrys sol*, the cysts cannot withstand desiccation. Helicoidal construction seems a good way to build a spherical shell, so as to withstand forces equally from any direction. This should enable it to cope with the forces of shrinkage during shrivelling, and of swelling during rehydration. For fuller details of spherical shells with helicoidal structure, see the relevant sections: protistans (Section 2.3a), nematodes (Section 2.3d), brine

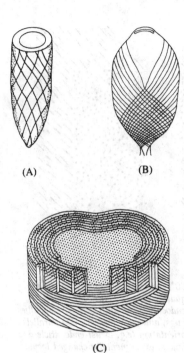

Figure 3.14. Some collagen fibre orientations in a squid, a tadpole, and a human intervertebral disc. (A) Diagram of the collagen fibres in the outer tunic enclosing the mantle muscles of the posterior part of a squid (Mollusca: Decapoda). The collagen forms helices of six to ten alternating senses that twist around the mantle. Redrawn from Ward and Wainwright (1972). (B) Diagram of the orientations of some of the collagen fibres in the basement membrane of a tadpole epidermis (Bombinator; Amphibia). Each layer is asymmetrical with respect to the whole body, but superposition of two layers gives the impression of pseudo-symmetry. Redrawn from Rosin (1946). (C) A diagram of the lateral walls of a human intervertebral disc from the lumbar region of the spine. Lamellae of collagen fibre plies are all set at 65° to the spine axis (here running vertically up the page). The fibre directions of the plies alternate from left to right (shown in cutaway view). The anterior and posterior ends of a disc are sealed by vertebral end plates, one of which is here shown stippled. Redrawn from Bogduk and Twomey (1987).

shrimp (Section 2.3i), and fish eggshells (Section 2.3p). Plant examples are included in Table 2.7, and moth eggshells are listed separately in Table 2.4.

3.2i Strategic designs I: insect exoskeletons

The use of fibrous composites in strategies designed to meet various mechanical criteria is well shown by examples from insect exoskeletons. Because of the established link between the appearances of insect cuticle in polarizing light and in the electron microscope (Fig. 1.20), it is possible to examine a whole region of cuticle in a polarizing microscope and interpret it as if the much more time-consuming transmission electron microscope (TEM) had been used. Each lamella seen in polarized light is equivalent to a single rank of arcs in the TEM, which itself corresponds to 180° rotation of helicoidally arranged chitin crystallites. In all of the examples there is a constant sense of helicoidal rotation. It is likely that areas of weakness would occur if rotation changed sense from region to region, or from time to time.

The mechanical design of insect cuticles may be understood by reference to Figure 3.15. These four types of chitin architecture derive from the two basic types – helicoidal or parallel – of our two-system model (Neville & Luke, 1969b). Figure 3.15A is exclusively helicoidal throughout its thickness. It is characteristic of exocuticles, which are all formed in the pharate stage of each instar before emergence takes place at the moult. Exocuticle is the most peripheral of the cuticular layers. This bears the brunt of stresses on the exoskeleton and will cope with them irrespective of direction. This architecture is also found throughout the endocuticle of endopterygote larvae (moth caterpillars, fly maggots, beetle grubs). It is appropriate for resisting the pressure of the blood equally from all directions, since these larval stages have non-rigid hydrostatic skeletons.

The cuticle in Figure 3.15B consists of an alternation of helicoidal and parallel layers. Typical examples are the jumping hind legs and the veins in the wings in locusts. During flight, wing veins need to counteract large bending forces; layers of chitin oriented parallel with the axis of a wing vein are well adapted for this. But if all the chitin were oriented in parallel with a leg segment or a wing vein, these would be liable to buckle when subject to twisting forces during flight. The alternate layers of helicoidal chitin provide stiffness and strength in all directions, some of which will resist shear and compression forces. It has been suggested that this helicoidal/parallel type of cuticle, which is confined to winged insects (Pterygota), may have evolved originally as an adaptation to flight (Neville, 1984). For a detailed study of wing structure in orthopteran grasshoppers, see Banerjee (1988a,b).

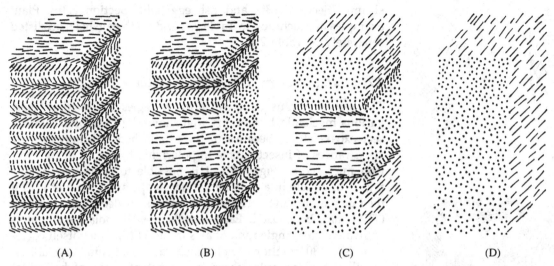

(A) (B) (C) (D)

Figure 3.15. *Types of chitin orientation in insect cuticles. Insect cuticles consist of combinations of two types of chitin architecture: helicoidal and parallel (Neville & Luke, 1969b). Each part of each insect exoskeleton has specific and precise structure. (A) Helicoidal throughout its thickness (e.g. exocuticles, endopterygote larvae). (B) Alternation of parallel and helicoidal layers, with all the parallel layers oriented in the same direction (e.g. locust endocuticle). There are usually several turns of helicoid per night-time layer, alternating with parallel layers formed in the daytime. (C) Pseudo-orthogonal, with parallel layers oriented at an angle of about 90°, connected by approximately a quarter-turn of helicoid (e.g. endocuticles of adult beetles, dragonflies, bugs, and stick insects). (D) Parallel orientation (e.g. locust endocuticle day layers, experimentally changed locust helicoidal layers, tendons).*

Figure 3.15C illustrates pseudo-orthogonal cuticle, typical of the endocuticle in beetles (adults) and of larval and adult bugs, stick insects, and dragonflies. In those few beetles which have been investigated, we know that the orientations of the major (parallel) layers are positioned with great precision (Zelazny & Neville, 1972). We do not yet know how these precise orientations meet the different requirements of the parts of a whole insect skeleton.

In the cuticle drawn in Figure 3.15D the chitin crystallites all run in parallel. This type is found in muscle tendons, where the forces predictably always act in a constant direction in series with the tendon. It is also found as a component of the type of cuticle in Figure 3.15B. Parallel chitin orientation is typical of hairs and spines and of ovipositors, which drill timber to lay their eggs (e.g. the wood wasp *Sirex* and its ichneumonoid parasitoid *Rhyssa*). In all these cases the chitin runs parallel to the long axis of the structure. The same is true for the piercing and sucking mouthparts of toe-biting water bugs (Belostomatidae) and for the stylets of aphids. The latter avoid penetrating plant epidermal walls (themselves helicoidal) by probing between epidermal cells, aided by pectinase enzymes.

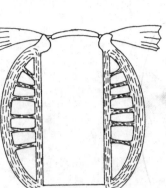

Figure 3.16. Strategic chitin orientations in a dragonfly thorax. Diagrammatical transverse section of a dragonfly thorax to show the scalariform apodeme on each side wall of the thorax. The predominant chitin microfibrillar orientations in these structures are shown by short lines. The ladder-like structures, which are found in the best fliers, are well designed to withstand the deformations caused by the thoracic flight muscles. Redrawn from Neville (1967c).

Moving to larger structural components, the cuticle in the flat swimming legs of giant water bugs (Belostomatidae) is dowelled together (Fig. 3.4) by pegs of cuticle, within which the chitin runs predominantly parallel with the axis of the peg (Neville, 1967c). The upper and lower surfaces of beetle elytra are joined together in a similar way. In the thorax in dragonflies, which houses their powerful flight muscles, the lateral walls are strengthened by ladder-like structures (scalariform apodemes). The predominant chitin orientations are arranged strategically as in Figure 3.16. This superb example of adaptation is designed to withstand the deformations caused by the flight muscles and is developed only in those kinds of dragonflies which are the best fliers.

3.2j Strategic designs II: human intervertebral discs

Human lumbar discs are situated between the vertebrae of the spine and are excellent examples of fibre orientation strategy. Their design and function are clearly described by Bogduk and Twomey (1987). Ten or twelve lamellar plies of collagen fibres form the lateral walls of each disc, enclosing a volume of hydraulic fluid. The two ends of a disc are sealed by vertebral end plates (shown stippled in Fig. 3.14C). The collagen fibres of each ply lie in parallel, are geodetic (i.e. they take any strain immediately), and are all set at 65° to the axis of the spine. Their sense of twist, however, alternates, so that odd- and even-numbered plies slant in opposite directions.

Pressure is transmitted from the incompressible fluid to the collagen fibres. When they are supporting the weight of the body, the discs are compressed, and all the fibres share the tensile strain resulting from the slight outward bulging of the side walls of each disc. During stretching of the spine (e.g. when hanging by the arms, or when in artificial traction), all the collagen fibres again share the tensile strain. But during mutual sliding or twisting of neighbouring vertebrae, because of the alternate directions of the plies, only half the fibres take the strain. (Half the fibres can resist a clockwise twist, and the other half an anti-clockwise twist). The significance of having fibres which run obliquely to the axis of the vertebral column is that every fibre has a component of resistance to either vertical or horizontal forces. Had the fibres been parallel to the spinal axis, they could carry no strain during either twisting or weight bearing. The angle of 65° is optimal for coping with the *variety* of forces. If the angle were steeper (less than 65°) it would benefit lengthening of the spine, but at the expense of being less able to cope with sliding or twisting forces. The opposite would apply for fibre angles greater than 65°.

Figure 3.17. Diagrams of the strategic orientations of fibres in three different systems. (A) Orientations of trabeculae in a sectioned human femur (Homo sapiens). These coincide with the directions of tension and compression, which cross each other orthogonally. The bone is well designed. Redrawn from Thompson (1961). (B) Major orientations of chitin fibres in the endocuticle of a beetle femur (Heliocopris colossus). The fibres form equiangular alternating left- and right-handed helices. The directions of maximum birefringence are indicated for two layers of chitin by biaxial indicatrices (ellipses). The fibres lie mainly at 45° to the femoral long axis (dashed line). In this way they will counteract maximum shearing forces arising from angular distortions. The beetle femur is also well designed. Redrawn from Neville (1967c). (C) The mechanical strategy of collagen fibres in the middle layers of the human cornea. These are described as being laminate and are arranged in criss-cross manner, with some interlacing. Towards the lateral (L) and median (M) edges, the collagen fibres become oriented predominantly in parallel, forming tendons to the rectus muscles. They thus serve an optical function within the cornea, and a mechanical function at its edge. Redrawn from Duke-Elder and Wybar (1961).

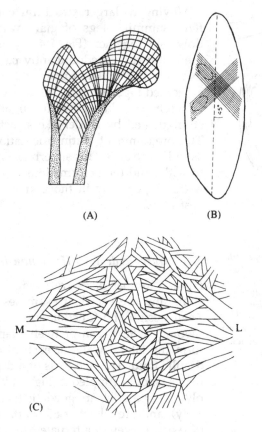

3.2k Strategic designs III: human bone

Bone provides another excellent example of strategic fibre design (Thompson, 1961). In the human femur, the fibres follow the predominant stress lines of tension and compression (Fig. 3.17A). The bone is thus optimally strong. Likewise, a beetle femur is well designed (Neville, 1967c), in this case to counteract maximum shearing forces operating at 45° to the femoral long axis (Fig. 3.17B). The major chitin fibre directions in the endocuticle are oriented similarly. The femur resists angular distortions.

3.2l Strategic designs IV: human cornea

The primary function of collagen fibres in the human cornea is to provide transparency. However, at the edges of the cornea they become predominantly parallel to the directions of the rectus muscles (Fig. 3.17C) and so act as tendons. According to a leading textbook on human corneal structure (Duke-Elder & Wybar, 1961), the collagen fibre directions vary according to their depth of position. In the earlier and deeper layers the fibres tend towards a circular orientation around the periphery. In the middle layers the fibres become oriented towards the insertions of the rectus muscles, while in the younger, more superficial layers the fibres tend to run vertically up and down. The corneal architecture in humans is thus more complicated than the twisted orthogonal structure of lower vertebrate classes. However, in view of what we now know about helicoidal structure, human cornea could perhaps repay reinvestigation?

3.2m Strategic designs V: arthropod cuticle moulting fracture planes

In addition to lines of strength, arthropods also require localized lines of weakness (fracture planes). This ensures that the exoskeleton will split in predetermined places during moulting. Little ultra-structural work has been done on fracture planes. In the insect order of parasitoids known as Strepsiptera, the puparium protrudes from the cuticle of the still-living host. As in some flies (Diptera), an eversible structure on the head (ptilinum) pushes off a cap to allow the adult parasitoid to emerge. Electron micrographs show that the chitin crystallites in the exocuticle run in parallel around the circular edge of the cap (Fig. 3.18). The puparium splits in a fibrous manner along this edge (Kathirithamby et al., 1990). This is a strategic adaptation, because parallel-oriented chitin is extremely rare in exocuticles. The chitin crystallites at this plane of fracture have a larger diameter (10–15 nm) than the average for cuticles in general (2.8 nm) (Neville et al., 1976). It is to be expected that the larger crystallites would give a weaker fibrous composite than would thinner crystallites.

3.3 Optical properties

Fibrous composite materials have evolved primarily to solve mechanical problems. They do, however, also have characteristic optical properties.

3.3a Birefringence

A substance is said to be birefringent if it presents different refractive indices to light rays travelling through it in different directions. A ray is slowed down by a medium of higher refractive index. Each ray may be considered to have a magnetic vector and an electric vector oscillating sinusoidally, in planes set mutually at right angles. When one vector travels in the direction of least refractive index and the other in the direction of highest refractive index, they will experience maximum birefringence. Using white light, this may be recognized as interference colours seen between crossed polarizing filters. [For a fuller description, including practical details, see Neville (1980).]

106 / *Properties of natural plywoods*

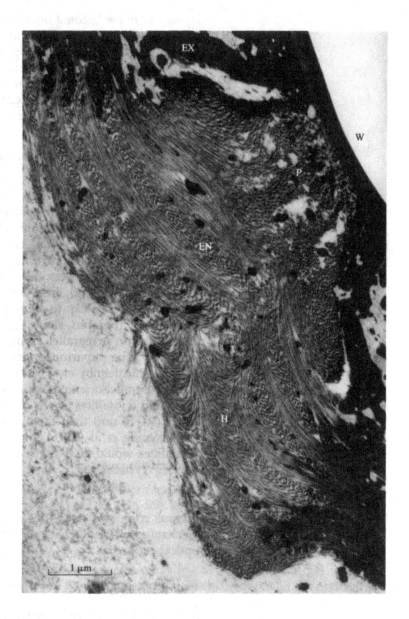

Figure 3.18. Chitin orientation and moulting in an insect. An electron micrograph to show a specific region in the cuticle of an insect where splitting will take place at the next moult. The cuticle architecture seems to be deliberately weakened in this region (W). The hard dark exocuticle (EX) is replaced at the predetermined line of weakness by mechanically weaker endocuticle (EN), whose chitin microfibrils run parallel to the direction in which they will ultimately split (P). The helicoidal endocuticle (H) at the breakage plane is looser (larger pitch) than in surrounding regions and has chitin microfibrils with more than ten times (33 nm) the normal diameter; this will give a weaker composite than usual. The material is from the puparium of an insect parasitoid (Elenchus tenuicornis) which belongs to the unusual order Strepsiptera. The larva is a parasitoid inside the larva of a bug (Javesella dubia; Homoptera, Delphacidae). As in higher Diptera (to which – rather than the beetles – they are now thought to be closely related), the cuticle of the final larva of Elenchus hardens into a puparium. The puparium protrudes through a hole in the cuticle of its host. To emerge, the adult parasitoid uses an inflatable ptilinum to break a neat round cap off its puparium (again like higher Diptera). This micrograph is of a section through the predetermined circular site of breakage in a male puparium. From Kathirithamby et al. (1990), with permission of Academic Press Limited, London, and of the authors.

In biological materials there are three possible kinds of birefringence: strain, intrinsic, and form. *Strain birefringence* is seen when a rubber (e.g. resilin) is observed between crossed polarizing filters. Initially, with the rubber under no strain, no interference colours are seen. But as the sample is stretched, the molecular chains begin to line up and so create a different refractive index in the direction of strain. Interference colours will then appear.

Intrinsic birefringence arises in those substances which form crystals with different refactive indices in different directions within the crystal's atomic lattice. This is readily seen in, for instance, the crystals of calcium oxalate in certain plant cells.

The most common type of birefringence in biological systems is *form birefringence*. The criteria for this type are met by a fibrous composite (such as the one illustrated in Fig. 1.3C) as follows: (1) It is a Wiener composite body with rods set in a matrix. (2) The rods and matrix must have distinct phase boundaries. (3) Both the diameter and spacing of the rods must be small compared to the wavelengths of visible light. (4) The rods should be lined up in some particular plane or direction. (5) Form birefringence should vary according to the refractive index of the matrix, which can be experimentally varied. If insect cuticle is penetrated by Canada balsam (a coniferous resin used for mounting permanent microscope slides), its form birefringence disappears, because it has the same refractive index as the chitin rods (1.54). By contrast, intrinsic birefringence cannot be altered by immersion in fluids of various refractive indices; the fluids are unable to penetrate the crystalline atomic lattice without causing the crystalline substance to dissolve. A mathematical formula for form birefringence has been published (Neville, 1980).

3.3b Helicoidal optics

Helicoids possess special optical properties. In oblique sections of systems with suitably large pitch (e.g. tunicate test or crab endocuticle), the unique property of *layer procession* may be observed (Fig. 3.19). For this, the polarizer and analyzer of the microscope must be rotated together. Whichever component layers in the helicoid are lying at 45° to the polarizers at any one time will appear brightly birefringent, whereas those layers then lying in parallel with either polarizer will appear dark. The dark and light bands will process through planes normal to the optic axis as the polarizers are rotated (Fig. 3.19). By careful interpretation, the sense of rotation of the helicoid may be deduced.

In helicoids of appropriate pitch (small with respect to the wavelengths of visible light), *interference colours* may be seen (e.g. in some cholesteric liquid crystals and some scarab beetle cuticles). For this, the optical system must be transparent, to permit penetration and subsequent reflection of incident light. Rays of suitable wavelength and direction reinforce each other by reflection from component layers situated 180° of helicoidal rotation apart. The two rays drawn as solid lines in Figure 3.20 will reinforce by constructive interference, as will those represented by the two dashed lines, provided they are viewed from a different (and suitable) direction. The colours seen will depend upon the spacing and regularity of the helicoid. The helicoid behaves as a half-wave interference reflector (Neville & Caveney, 1969). Colours complementary to those reflected will be transmitted

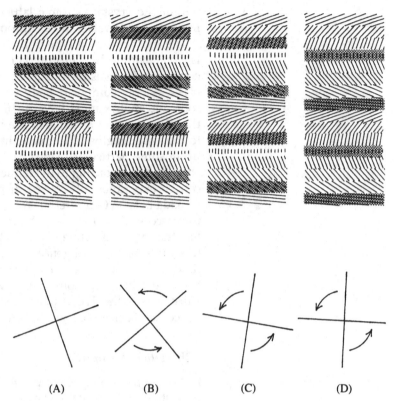

Figure 3.19. An optical method to detect helicoids. Given a system with a very large helicoidal pitch (e.g. up to 100 μm in a tunicate test), a unique optical property may be observed. The figure shows an oblique section through 360° of helicoid, as viewed between crossed polaroids. All planes of microfibrils which run parallel to the polaroid axes appear as a dark band (shown stippled). If the polaroids are rotated (as a yoked pair) as shown by the arrows, in a sequence from part A to part D, there will be a procession of dark bands. The direction of procession depends upon the sense of a helicoid. This method may be used to determine the sense of a helicoid even if the pitch is too small for resolution of arced patterning.

through the helicoid. For example, if green is reflected, then blue and red will be transmitted and will add to appear purple. A helicoid of constant pitch reflects a narrow waveband of pure saturated colour, whereas one of varying pitch may reflect all wavelengths. If a large number of helicoidal turns is involved, then a silver-mirror (broad-band) appearance will result (Neville, 1977). As many as ninety-five half-turns of helicoid occur in the exocuticle of the golden beetle, *Plusiotis resplendens* (Caveney, 1971).

Another unique optical property of helicoids is the ability to reflect or transmit *circularly polarized light* of opposite senses. This is found in association with the reflection of interference colours, described in the preceding paragraph. These unusual optical properties are described with reference to Figure 3.21. Ordinary light (O) strikes a transparent helicoidal stack of layers (H). If the pitch of the helicoid is of appropriate size, reinforcement will take place by constructive interference (as described earlier), and light rays of a particular wavelength will be *reflected*. We see this as a particular colour of our visible spectrum. Other (complementary) wavelengths are *transmitted* through the helicoid. The

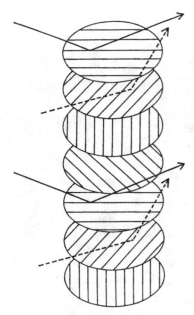

Figure 3.20. How helicoids may produce interference colours. The ellipses represent directions of chitin microfibrils in a helicoid. If the pitch is appropriate, rays of a specified wavelength (and direction) reinforce each other to produce a pure spectral colour. For example, in the diagram, the two rays drawn in solid lines will reinforce each other, as will the two drawn in dashes.

helicoid (H) drawn in Figure 3.21 is left-handed; the convention (adopted from the field of liquid crystals) is that in a *left*-handed stack the orientation of the individual layers turns anti-clockwise in layers situated progressively *farther away* from the observer.

Rays reflected from a left-handed stack are left circularly polarized: The electric (hatched) and magnetic (dotted) vectors vibrate sinusoidally and in planes which are mutually at right angles, and a quarter of a wavelength (90°) out of phase with one another. The resultant of these two sine waves plotted in three dimensions is a Riemann surface which (theoretically) spirals *towards* the observer. Notionally this oncoming spiral would appear circular; hence the name circularly polarized light. The convention for circularly polarized light is that it is *left*-handed if the spiral *approaching* the observer appears to rotate anti-clockwise; it should be noted that the conventions for helicoidal stacks and circularly polarized light are 180° out of phase with each other. The reflected circularly polarized ray (L) in Figure 3.21 is left-handed, whereas the transmitted ray (R) is right-handed. Had the helicoidal stack been right-handed, the reflected ray would have been right circularly polarized, and the transmitted ray left circularly polarized. (In the more commonly encountered *plane* polarized light the two vectors also vibrate in planes set mutually at right angles, but they are *not* out of phase. Their resultant is a further sine wave oscillating in a plane at 45° to those of the two vectors.)

By making use of suitable optical accessories, the sense of rotation of the circularly polarized light, and hence also that of the helicoidal stack of layers which produced it, may be determined. [For practical details, see Robinson (1966), Neville & Caveney (1969), and Neville (1980).] This observation is easily performed. We took advantage of this piece of luck to check several thousands of species of scarab beetles in the collections of the British Museum of Natural History in London (Neville & Caveney, 1969). For those which reflected circularly polarized light, the sense was always left-handed, as drawn in Figure 3.21. We may deduce that they all have left-handed helicoidal ultra-structure in their cuticles.

Like scarab beetles, cholesteric liquid crystals of appropriate pitch also display interference colours and reflection of circularly polarized light. They also show extinction of the colours [a photograph of which was published earlier (Neville, 1980)] when viewed through a right circular analyzer (a combination of a quarter-wave sheet of plastic overlain by a polarizing sheet). Left circularly polarized light reverses through 180° when viewed in a mirror, so that it appears as right circularly polarized light. This is because the magnetic vector now leads the electric vector, rather than vice versa. Observed through a right circular analyzer, a beetle appears black, but its reflection is still coloured. The

110 / *Properties of natural plywoods*

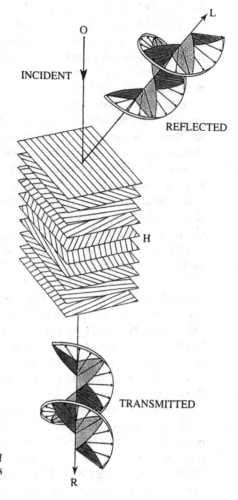

Figure 3.21. The interrelation between a helicoid and circularly polarized light. A three-dimensional representation of an incident ray of ordinary white light (O) striking a transparent left-handed helicoidal stack of layers (H). Rays of suitable wavelength are reflected as left circularly polarized light (L), whereas other rays are transmitted as right circularly polarized light (R). By convention, a structure is denoted as left-handed if its layers rotate anti-clockwise progressively farther away from the observer – like walking down a left-handed spiral staircase. By contrast, circularly polarized light is called left-handed if the resultant spiral rotates anti-clockwise as it approaches the observer.

appearances are reversed if a left circular analyzer is used. Cholesteric liquid crystals do the same.

That these circularly polarized interference colours arise in a helicoidal system was convincingly shown by Mauguin (1911) using liquid crystals. He assembled nematic (parallel) liquid crystals of azoxyanisol and azoxyphenetol between a microscope slide and cover-slip. When the cover-slip was mechanically rotated through many turns (always with the same sense of torsion), interference colours eventually appeared. These show the same optical properties as scarab beetle exocuticle and self-assembling cholesteric liquid crystals. The similarity of optical properties in scarabs and liquid crystals is a strong indicator that natural helicoids could arise from self-assembling liquid crystals. This idea is pursued in Chapter 4. Mauguin's experiment is important because it makes the helicoidally twisted concept tangible.

Another special property of helicoids is that they show two

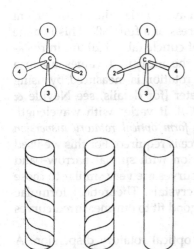

Figure 3.22 (above). Diagrams to show the two mirror-imaged ways in which four different chemical groups may attach to an asymmetrical carbon atom. This asymmetry may, together with Brownian motion, give rise to the twisting force which drives the assembly of helicoids.

Figure 3.23 (below). Fibrils of polybenzylglutamate as observed in the electron microscope. (A) The laevo-rotatory form (PBLG) forms D α-helices which twist to form the left-handed fibril drawn here. (B) The dextro-rotatory form (PBDG) forms L α-helices which twist to form the right-handed fibril drawn here. (C) A racemic mixture (equal amounts of PBLG and PBDG) cancels the twists, resulting in non-helical fibrils. Based on results by Tachibana and Kambara (1967).

kinds of *optical activity*. An optically active material is capable of rotating the plane of a beam of plane polarized light that is transmitted through it. The normal kind is *molecular* optical rotation, which originates in asymmetric carbon atoms. An *asymmetrical carbon* is bonded to four different groups (Fig. 3.22) and hence can exist in either of two mirror-imaged forms (L or D). These rotate plane polarized light to the left or right, respectively. Naturally occurring amino acids are always in the L-form, whereas sugars and amino-sugars (e.g. in chitin) are in the D-form. L and D isomers form the basis of the sense of twist in polymer microfibrils (Fig. 3.23).

It is the second type of optical activity which is special to helicoids; this arises in the helicoidal array of chitin crystallites (i.e. at the fibrous composite level). By analogy with intrinsic birefringence and form birefringence (the latter also arising at the fibrous composite level), we named this activity *form* optical rotation (Neville & Caveney, 1969), so as to distinguish it from molecular optical activity. Form optical activity occurs in combination with the reflection of circularly polarized light, found in the cuticular helicoids of scarab beetles and in the cuticle of a few crustaceans (Neville & Luke, 1971a). The major difference between the two types of optical activity is that the angular rotation imparted to plane polarized light by a 1-mm-thick sample of form optically active beetle cuticle is about an order of magnitude greater than that of a similar thickness of concentrated solution of a molecule which shows only molecular optical rotation. Typical values are 1,000° and 100° per millimetre, respectively.

One special beetle shows form optical activity as high as 10,000° per millimetre thickness of exocuticular helicoid (Caveney, 1971). This is a brilliantly iridescent golden beetle (*Plusiotis resplendens*) which deposits tiny crystals of uric acid in its exocuticle. The uric acid may amount to as much as 70% of the exocuticular weight. As this is a desert beetle, this is probably an adaptation for excreting uric acid whilst conserving water. This huge value of form optical rotation indicates that the crystallites of uric acid are oriented in the same helicoidal array as the chitin crystallites – probably by epitactic crystallization. The uric acid crystals must be in the needle form (the same form which gives the pain suffered in gout when they prick nerves), rather than the plate form. Samples of exocuticle of *Plusiotis resplendens* lose their iridescence when soaked in ethanol. When Stanley Caveney made this observation in my laboratory in Oxford, we were initially very worried – we had just completed our work on scarab beetle optics (Neville & Caveney, 1969). Physical interference colours (on which the whole analysis of circularly polarized light and optical activity depends) are not supposed to dissolve! However, the problem was solved when Caveney went on to show that the ethanol had dissolved out the uric acid.

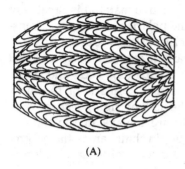

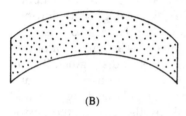

Figure 3.24. Normal and mutant compound eye lenses in Drosophila melanogaster. *(A) A normal lens has biconvex shape, is thicker (5.6 μm), is transparent, and has up to eleven half-turns of continuous helicoidal chitin architecture (shown by arced patterning). (B) A mutant lens has convex outer and concave inner surfaces, is thinner (2 μm), is opaque, and has chitin crystallites in parallel (represented as dots in section). Based on results of Youssef and Gardner (1975).*

The astute reader will by now have queried the measurement of angles of optical rotation in excess of a full 360°. This is done by preparing very thin samples of cuticle (about 20 μm in thickness) and linearly extrapolating the measured values up to a thickness of 1 mm. Form optical rotation is measured by using a light microscope as a polarimeter [for details, see Neville & Caveney (1969) and Neville (1980)]. It varies with wavelength (dispersion), so that to record a *form optical rotatory dispersion* curve a monochromatic light source is required. For this we used a white light source in combination with special narrow-band interference filters. The resulting curves are very similar to those measured for cholesteric liquid crystals. Theoretical formulae derived for liquid crystals gave a good fit to our measured curves (Neville & Caveney, 1969).

There is a quick test for form optical rotatory dispersion. A thin fragment of exocuticle from an iridescent scarab beetle is placed between crossed polarizers and observed in a light microscope. A bright colour corresponding to the colour of the beetle is seen. On rotation of the specimen, no brightness maxima or minima occur; this distinguishes the phenomenon from that of birefringence, which passes through four maxima and four minima of brightness in a single rotation of the specimen.

A surprising property of optically active scarab exocuticle was noted in our review (Neville & Caveney, 1969): The birefringence of sections through the exocuticle is positive in a direction perpendicular to the cuticle surface. This 'anomalous' birefringence is reinforced if the sections are stained dichroically with toluidine blue. Furthermore, the mechanical cleavage direction is also perpendicular to the cuticle surface. If the protein is chemically removed, the direction of birefringence reverses to become positive parallel to the cuticle surface. In my experience, all other types of insect cuticle show birefringence in parallel with the chitin crystallites. The foregoing observations remain unresolved; perhaps they indicate that the orientation of the protein chains is predominantly along the helicoidal axis of rotation.

In conclusion, the optical properties described earlier provide strong support for the helicoidal interpretation of insect cuticle. They also show great similarity to the cholesteric liquid crystalline state discussed in more detail in Chapter 4.

3.3c Insect eye lenses

In the fruit fly *Drosophila melanogaster* there is a mutant (Fig. 3.24) with opaque compound eye lenses (Youssef & Gardner, 1975). A normal wild-type fruit fly has eye lenses which are bi-convex, about 5.5 μm in thickness, transparent, and with up to twenty-two half-turns of helicoidal chitin. By contrast, the mutant has lenses which are convex on the outside but concave on the inner

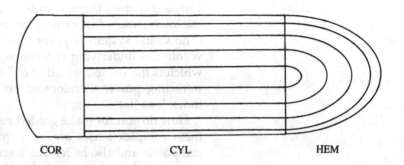

Figure 3.25. Diagram of the parts of a graded refractive index cuticular lens from a scarab beetle, drawn in longitudinal section. The lens has three regions: cornea (COR), graded refractive index cylinder (CYL), and graded refractive index lens hemisphere (HEM). The lines are contours of equal refractive indices. Redrawn from Caveney and McIntyre (1981).

surface, only 2.0 μm in thickness, opaque, and with parallel-oriented chitin (despite the ability to produce helicoidal chitin in the surrounding cuticle). A small number of facets had one to four half-turns of helicoid in the mutant, but most had unidirectional chitin throughout. This material seems to demand further work. Is the mutant gene normally concerned with producing a protein connected with helicoidal cuticle deposition?

The cuticle of compound eye lenses differs from that of ocellar lenses in general in lacking pore canals. These would presumably interfere with image production. By contrast, ocelli do not produce images – they merely detect light intensity – and they do have pore canals running through the lens in the same direction as the axes of helicoidal rotation.

It has been shown that in the majority of a sample of forty species of scarab beetles which were tested, the compound eyes have *graded refractive index lenses* (Caveney & McIntyre, 1981). This is surprising, because curved lenses with graded refractive indices are difficult to manufacture. A diagram of a longitudinal section through a graded refractive index cylindrical corneal lens facet from a scarab beetle is shown in Figure 3.25. The corneal facet may be considered in three parts: the outer part of the corneal facet, the graded refractive index cylinder, and the graded refractive index lens hemisphere. The outer part of the corneal facet has a uniform refractive index in all directions; it is isotropic. In the other two parts of the lens the refractive index varies with the radius, with the highest value along the central axis. The lines drawn on the cylinder and hemisphere parts of the diagram represent contour lines with the same refractive index (RI). The RI decreases from the central optic axis to the outside surface.

Caveney and McIntyre (1981) provide a thorough analysis of the properties of these complex lenses, including ray tracing through the optical system. In terrestrial vertebrates, the refracting power of the eye is provided by the outer curvature of the cornea, with very little contribution from the lens, whose function is to control accommodation. In scarab beetles, however, the cornea is so thick that if its outer surface had too large a curvature, the intermediate focus would occur within the corneal facet itself. (The visual system requires that the intermediate focus occurs within the underlying crystalline cone – a separate component which is the pooled secretion of four cells.) In scarabs, the major refracting power therefore occurs within the graded refractive index lens itself.

How do scarabs make graded refractive index lenses? Because these complex lenses are built from cuticle which is a fibrous composite, and also helicoidal, it seems reasonable to ask whether a controlled gradation in helicoidal pitch might be involved in the production of the graded refractive index effect.

3.3d Transparency in bird cornea

Maurice (1957) considers that transparency in corneal tissue depends upon properties of the collagen fibrils. These should be of small and uniform diameter and spacing, with orthogonal arrangement. Opacity may result if control of collagen fibril diameter goes wrong (Anseth, 1965). An orthogonal array of collagen fibrils acts as a diffraction grating (Maurice, 1957). The main beam of light passes through a diffraction grating undiffracted. Off-centre rays (scattering) will obey the formula

$$n\lambda = d \sin \alpha$$

where n is the order of diffraction, λ is light wavelength, d is collagen fibril spacing, and α is the angle of a diffracted ray. Diffracted rays will occur only if $n = 0$, and will be lost by destructive interference. Only the main beam is transmitted, and no scattering spoils the transparency. (The cells within the vertebrate cornea do in fact cause about 2% loss of light by scattering. This is a problem not encountered by arthropods, whose cornea is non-cellular.)

The cornea of the embryonic chick is originally opaque because it contains hyaluronic acid. This binds a large amount of water, which swells the cornea, so that the spacing of collagen fibrils (although orthogonal in array) is so great that they scatter light. At a later stage of development the enzyme hyaluronidase is secreted; this breaks down the hyaluronic acid, releasing its bound water, which is actively pumped out of the cornea by its endothelium. The collagen fibrils move closer together and establish transparency (Bard, 1990). We may note that even if a cornea has collagen fibrils spaced so closely as to cause iridescence (e.g. in the fish *Zeus*), it may still be transparent. There is a need for a comparative survey by electron microscopy of bird corneal collagen architecture. It has been suggested (Woodhead-Galloway, 1980) that precision of arrangement of corneal collagen

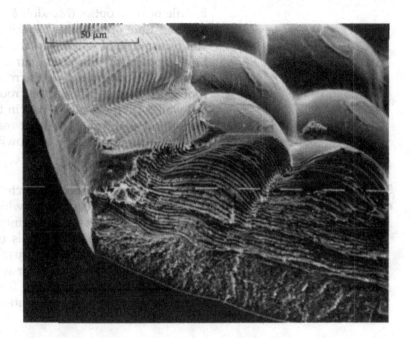

Figure 3.26. Scanning electron micrograph of a broken portion of a few facets from the compound eye of a giant Malaysian stick insect (Heteropteryx dilatata). *The lamination arises in a helicoid; arcs cannot be resolved in this material by this method. From Loder (1989), by kind permission.*

could aid acuity of vision in those birds whose feeding habits most require it.

The helicoidally twisted orthogonal collagen architecture in bird cornea is ideally suited to resisting intraocular pressure equally from all directions. A spherical shape is established and maintained, with minimum spherical aberration. Because the cornea forms the main refractory surface in a bird's eye, its helicoidal structure is important.

3.3e Insect and bird corneas and navigation

The individual corneal lenses which make up an insect compound eye are helicoidally constructed (Fig. 3.26). The sense of helicoidal rotation is the same in both compound eyes (Neville, 1976). This means that, as in those birds which have been thoroughly investigated (Trelstad, 1982), the fibrillar construction of left and right eyes in insects is also asymmetrical with respect to the bilateral symmetry of the body. Could this have any significance in relation to navigation and homing abilities, perhaps by the use of polarized light from the sky? Pigeons and honey bees would appear to warrant investigation. It is the outermost layers of bird cornea which have rotated structure, and the outermost layers of insect eyes have the smallest pitch.

3.3f Plant fibre optics

It has recently been demonstrated that the columns of cells of plant seedlings (e.g. oats, corn, and mung bean) may act like a bundle of fibre optics (Mandoli & Briggs, 1984). As with manufactured fibre optics, laser experiments show that seedlings can still transmit light when they are bent. As with fibre optics, the light zig-zags off the inside surface of the fibre and can even penetrate below ground via the roots. Light travels to the tips of root hairs. Is it significant that root hairs often have helicoidal cell walls? Do these play a part in the total internal reflection of the optical pathway? The light may act to coordinate various aspects of the physiology of growing plants.

3.4 Influences of fibrous composites on cells

That cells may influence the architecture and properties of the extracellular fibrous composites which they secrete is self-evident. The reciprocal effect – that seemingly static cuticles, basement membranes, and plant cell walls may have dynamic influence over nearby cells – is less obvious. This section gathers together diverse examples of extracellular systems which affect cells; they include arthropod exoskeleton, fish eggshells, basement membrane, vertebrate cornea, vertebrate extracellular matrix, and plant cell walls.

3.4a Pore canal shape in arthropod cuticle

Pore canals are numerous fine extensions of the epidermal cells in arthropods. They traverse the exoskeletal cuticle, carrying waxes to its outer surface for repair of any scratches in the waterproofing outermost layer. It is not surprising to find that these canals are flattened in the direction of the surrounding chitin crystallites (Neville et al., 1969). Each pore canal is shaped like a flat straight ribbon when traversing a layer of parallel chitin crystallites, but like a flat twisted ribbon (Fig. 3.27A) when passing through a layer of helicoidal chitin (Neville & Luke, 1969b). The sense of twist is determined by (and therefore is the same as) the sense of the helicoid (i.e. left-handed). The shape of larger ducts (e.g. dermal gland ducts and sensory ducts) is not affected by the chitin architecture. Oblique sections of helicoidal cuticle show characteristic crescentic pore canal shapes which form a pattern coincident with the ranks of arcs created by the chitin crystallites (Fig. 3.27B). Both patterns repeat every 180° of helicoid. This coincidence of patterns led to the rejection of the helical model for pore canal shape in favour of the twisted ribbon shape, because when sectioned obliquely a helix creates a pattern which repeats not every 180° but every 360° (Neville et al., 1969).

The pore canals in a particular piece of cuticle twist in phase. Our twisted ribbon interpretation of pore canals in insects was confirmed for spider cuticle by Barth (1970), who was able to visualise twisted ribbons by light microscopy. Our interpretation has also been confirmed for the pore canals of a crab (*Carcinus*

3.4 Influences of fibrous composites on cells / 117

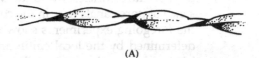

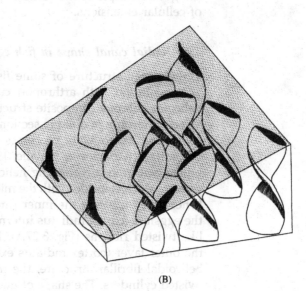

Figure 3.27. Twisted pore canals in insect cuticle. (A) The shape of an isolated pore canal from insect cuticle; it twists in phase with the surrounding chitin helicoid. From Grierson and Neville (1981), by permission of Churchill Livingstone Journals, Edinburgh, U.K. (B) Diagram showing the origin of crescentic patterns (in black) of pore canals arising on an oblique plane of section (stippled), cut through a field of canals which twist in unison. Canals are cellular extensions of the underlying epidermal cells, which traverse the cuticle. Reconstructed from a transmission electron micrograph of crayfish (Astacus fluviatilis) cuticle. From Neville and Berg (1971), by permission of the Palaeontological Association, London.

maenas) by Compère and Goffinet (1987), who refined our model by drawing sharply ended crescentic sections rather than our (diagrammatic) elliptical sections. [I had already published such a diagram (Neville, 1975a, Figure 7.6).]

We were able to produce sections of fossil (Jurassic) lobster cuticle (*Eryma*) which showed pore canal profiles like those of present-day *Astacus*. It was therefore possible to infer that helicoidal cuticle already existed in the age of dinosaurs (Neville & Berg, 1971).

It is possible to demonstrate that manipulation of the environment at the time of endocuticle deposition in locusts simultaneously alters both chitin architecture and pore canal shape. Locust cuticle normally contains parallel (daytime) and helicoidal (night-time) layers in alternation (Fig. 1.18). These are timed by circadian clocks (see Section 5.2h). Cuticle deposited during continuous day conditions (light and high temperature) is parallel throughout. By contrast, cuticle deposited during continuous night conditions (dark and lower temperature) is helicoidal throughout. Pore canal shape follows the prevailing chitin architecture. In control conditions, each pore canal will change alternately from twisted to untwisted along its length in a daily manner. In cuticle with continuously parallel chitin rods, the pore canals are entirely untwisted; in cuticle which is completely helicoidal, the pore canals are entirely twisted. We can separate cause and effect here because cuticles are found with helicoidal chitin in the absence of pore canals (e.g. in the compound eye cuticle). Therefore the foregoing experiments show that the shape of pore canals is determined by the local chitin architecture, not vice versa. This is a good example of extracellular products controlling the shape of cellular extensions.

3.4b Radial canal shape in fish eggshells

The eggshell structure of some fish shows remarkable convergent evolution with arthropod cuticle. Not only does it have helicoidal fibrous composite structure, yielding typical ranks of arced patterning in oblique sections, but it also has radial canals containing microvillate processes which arise from the oocyte cell surface (Grierson & Neville, 1981). These microvilli twist in unison with the surrounding helicoid and generate a crescentic pattern which coincides with the microfibrillar pattern (Fig. 3.27B). Where they traverse the inner (more recently secreted) part of the eggshell (cortex radiatus internus), the microvilli are shaped like twisted ribbons (Fig. 3.27A). But where they pass through the outer layer (cortex radiatus externus), which does not have helicoidal fibrillar structure, the microvilli are shaped like untwisted cylinders. The shape of microvilli is therefore cylindrical in the region farthest from the oocyte cell, and twisted ribbon in the region nearest the oocyte cell. Like pore canals in arthropod cuticle, their shape is determined by the surrounding extracellular fibrous composite.

3.4c Effect of locust cuticle on growth of fungal hyphae

Fungal pathogens of insects (e.g. *Beauveria* and *Metarrhizium*) invade the cuticle by secreting digestive enzymes such as lipases and chitinases. These enzymes follow the pore canal system, making way for hyphal penetration. The fungal hyphae follow the daily growth layer system in locust cuticle (David, 1967), penetrating along the parallel daytime layers which provide the least resistance.

3.4d Oligosaccharins: hormones from plant cell walls

Recent exciting new research is showing that plant cell walls exert dynamic control over the cells which have secreted them. Classical plant hormones such as auxins, abscisic acid, gibberellins, and ethylene are fairly unspecific in their effects. Each of them triggers a number of different events – and there are only a few different kinds. It is now emerging that fragments of hemicelluloses, released enzymatically from plant cell walls, may exert much more specific control. Auxin acts on the cell plasma membrane to release hydrogen ions (Rayle, 1973). These activate enzymes (e.g. glucanases) which release small fragments from branched hemicelluloses built into the cell wall matrix. The small molecules released are themselves branched, are known as oligosaccharins, and may typically contain nine residues (Albersheim & Darvill, 1985). It is their branched shape which endows them with specificity; the number of shape permutations possible by combining branching with just a few kinds of sugars is enormous.

Oligosaccharins work in extremely low concentrations (e.g. 10^{-8} to 10^{-9} M) to give specific control over growth, root formation, flowering, and so forth (Tran et al., 1985). It is possible that some oligosaccharins could act as feedback inhibitors of auxin-stimulated growth (York, Darvill, & Albersheim, 1984). Such regulatory control mechanisms are well known in animals, but this is the first demonstration of such a mechanism in plants. I have stuck out my neck (Neville, 1988a) and suggested that the complexity of branched hemicelluloses may have evolved as an adaptation to forming helicoidal cell walls by self-assembly (see Section 5.1). These could then have provided the source of oligosaccharin hormones of specific molecular shapes. A cascade control system, like that which controls the physiology of mammals, could then have arisen. The oligosaccharins are the plant equivalents of mammalian oligopeptides. In this context it is exciting to note that Vian et al. (1986) have recognized the involvement of fucogalactoxyloglucans in both nonosaccharide hormones and helicoidal self-assembly. Clearly, plant cell walls are not there simply to serve a skeletal purpose; they are decidedly active in plant physiology.

3.4e Contact guidance

This phrase was coined by Weiss (1961) to identify the process by which developing mobile cells follow specified trails in an embryo. Examples are known in which cells are guided by an extracellular trellis of collagen fibres. Several of the examples concern basement membranes. In fish, fibroblasts migrate from the lateral line region by following bundles of collagen fibrils which are oriented at 45° to the lateral line (Fitton-Jackson, 1968). During the development of the fins in teleost fish, collagen fibrils (called actinotrichia) guide migrating mesenchyme cells up into the fin (Wood & Thorogood, 1987). In amphibian tadpoles, pigmented cells (melanophores) take up positions on the collagen trellis of the epidermal basement lamella. The directions of cell processes reveal the orientation of the collagen (Rosin, 1946). This principle was also used by Overton (1979) to show that embryonic chick fibroblasts will line up on a collagen trellis

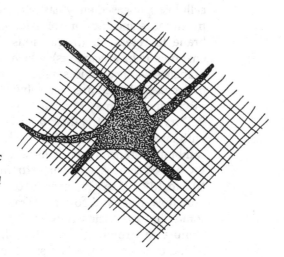

Figure 3.28. Collagen plywoods may influence cell shape. A chick embryonic fibroblast growing in cell culture on a collagen grid isolated from a larval tail of a clawed toad (Xenopus laevis). The fibroblast processes line up with the collagen fibre grid. Drawing based on a scanning electron micrograph by Overton (1979).

isolated from the tail of a *Xenopus laevis* tadpole (Fig. 3.28). This shows that the principle of contact guidance crosses phyletic boundaries, since an amphibian fibre pattern can influence the positioning of cellular processes of a bird. This property of collagen trellises is not merely of academic interest. First- and second-degree burns in humans can heal without leaving permanent scar tissue provided that only the Malpighian cells are wounded. If, however, the seemingly inert basement lamella of the cells is also damaged, as in third-degree burns, then only scar tissue can be formed. The collagen trellis is a vital structure.

Another collagenous system which shows contact guidance is the cornea of the vertebrate eye. This is ideal for experimentation because of its transparency. Migrating fibroblasts may be observed and filmed *in situ* (Hay, 1980). Using fowl chicks (*Gallus*), Bard and Higginson (1977) have demonstrated that collagen fibres must have undergone thickening from the primary stroma stage (0.4 µm in diameter) to the secondary stage (0.7 µm) before they are able to control the alignment of fibroblast cells which have invaded the cornea. Migrating fibroblast cells are ciliated and become flattened in the plane of the collagen lamination. When seen in section, the cornea eventually appears layered with fibroblast cells, which are in contact via tight junctions (Hay & Revel, 1969). In chick cornea the fibroblasts lay down additional collagen fibrils which follow the existing layer directions; these are orthogonal in the innermost layers, and helicoidal in the outermost layers.

3.4f Influence of vertebrate extracellular matrix on cells

The extracellular matrix (ECM) of vertebrates includes the composite fibre and matrix structure of connective tissues, cartilage, cornea, skin, blood vessel walls, intercellular fluid compartments, fluid in intervertebral discs, synovial fluid, and vitreous body. It represents a huge field of research, of interest to mainstream developmental biologists as well as to the medical community. Mention of ECM is here restricted to its effects on cells adjacent to it or within it. For an introduction to the literature, see Hay (1981), Arnott, Rees, and Morris (1983), Trelstad (1984), Bard (1990), and Adair and Mecham (1990). Two aspects of ECM are included in this book – basal lamina with basement lamella (Section 2.2f), and cornea (Sections 2.2h and 5.1d) – because their architecture is dominated by collagen fibres. Aspects of cell behaviour and development which are influenced importantly by the ECM include cell movement, shape, polarity, metabolism, and differentiation. A major difference between ECM and mechanical support systems in plant cell walls and insect cuticle is that cells (e.g. fibroblasts and mesenchyme) invade it and become isolated within it.

3.4g A helicoidal plant host–parasite interface

The fungus *Synchytrium mercurialis* is a parasite of the dog's mercury plant (*Mercurialis perennis*). The helicoidal chitinous sporocyte wall of the fungus moulds the cellular extensions of the host into twisted ribbon shapes. Oblique sections reveal a crescentic pattern like those described for insect cuticles and fish eggshells in Sections 3.4a and 3.4b. This interpretation is based upon the work of Meyer and Michler (1976). Thus, in this example the fibrillar system from one species influences cellular shape in another species. The interface acts as a haustorium to convey food from host to parasite.

In conclusion, fibrous composites have varied chemical compositions and physical properties and (despite being extracellular) profound influences on the cells which secreted them.

4
Biomimicry: making liquid crystalline models of helicoids and other plywoods

A chapter on liquid crystals is relevant to a book on fibrous composites for three main reasons: They both are mobile, they have similar architectures, and they both can self-assemble. The fibres of natural composites consist of large water-insoluble molecules which are put into place extracellularly. To be effective, these macromolecules must be strategically oriented (see Chapter 3). During development they therefore require to pass through a mobile phase, during which their positions can be adjusted, before solidification takes place. This chapter will show how liquid crystals meet this requirement.

The similarities between liquid crystals and insect cuticles were realized independently by Conmar Robinson (a retired polypeptide chemist) and myself and were presented in talks to a symposium on insect cuticle held in the Shell Building (London) in 1967. One presentation emphasized the optical similarities in the two systems (Robinson, 1966); the other compared their helicoidal structures and proposed that insect cuticular ultrastructure might develop by self-assembly via a liquid crystalline stage (Neville & Luke, 1969b). One of my main contributions has been subsequently to develop this idea in other systems, including plant cell walls, moth eggshells, fish eggshells, and mantis eggcases. The hope is that this will encourage biochemists to recognize features of structural chemistry which are relevant to liquid crystalline self-assembly.

A significant difference exists between biological composites and liquid crystals, despite their similar architectures. The fibrous component in composites consists of crystallites (small parallel bundles of molecules), whereas that of liquid crystals consists of individual molecular chains. The full significance of this difference is not yet understood (see, however, Section 4.3c). A further difference is that completed biological composites exist in the solid state, whereas liquid crystals exist in a separate state of matter all of their own. The solid composite counterparts of liquid crystals are known as analogues (Neville & Caveney, 1969) or (in French) as *pseudomorphoses* (Bouligand, 1969).

The term 'biomimicry' implies an attempt to copy biological structures synthetically; it is becoming fashionable. I have used it in the title of this chapter because it is about making models with liquid crystals. In fact, I almost coined it (in principle) by noting how the different regions of a β-keratin molecule 'mimic' the fibre and matrix components of a two-phase composite (Neville, 1975a, p. 263). Perhaps the term will help the subject matter of this book to become more respectable (and grantworthy?).

There are some superbly illustrated papers by Yves Bouligand and colleagues on liquid crystalline structures, and reference should be made to these for further details (Bouligand, 1969, 1981a,b; Bouligand & Kléman, 1970; Bouligand & Livolant, 1984).

This chapter introduces some principles of liquid crystals. It then shows how information from synthetic liquid crystalline materials may help us to understand the structure and assembly mechanisms of biological composites via a liquid crystalline phase. Some examples of naturally occurring liquid crystals are then presented. Developmental aspects are considered in more detail in Chapter 5.

4.1 Principles and types of liquid crystals

Liquid crystals represent a fourth state of matter, with properties intermediate between those of liquid and solid. (They are also known as mesophases or mesomorphic phases.) They combine the viscosity and flow of liquids with the long-range order of solids (e.g. they show birefringence and even simple X-ray diffraction patterns). They are capable of self-assembly into either simple or quite complex geometrical shapes. The molecules are usually elongated and fairly rigid. Several types of liquid crystals exist, and three of these are relevant to extracellular developmental biologists.

4.1a Nematic liquid crystals

In nematic liquid crystals (Greek *nema* is a thread), long rod-shaped molecules line up in parallel like logs in a stream. A better analogy, because it conveys a three-dimensional impression, is that the molecules align like a school of fish. This analogy also conveys the mobility of the molecules. When long rod-shaped molecules reach a critical concentration within a fluid, they automatically line up in this way for geometrical reasons. The architecture is like that of the diagram of chitin crystallites in Figure 1.3C.

4.1b Cholesteric liquid crystals

In cholesteric liquid crystals, long rod-shaped molecules have helicoidal distribution like that in the stereoscopic diagram in Figure 1.9. The structure is only apparently layered; in a very small volume of a cholesteric liquid crystal the molecules run almost in parallel. There is, however, a slight angular change between neighbouring molecules, always in one sense of direction for a specified set of conditions. The term 'cholesteric' is misleading, as the substance cholesterol does not form cholesteric liquid crystals. (Esters and ethers of cholesterol do, however, form cholesteric liquid crystals; e.g. cholesteryl benzoate, cholesteryl succinate.) Perhaps it is more satisfactory to call them 'twisted nematic' or 'chiral nematic' liquid crystals; it will be shown later that it is possible to convert a nematic into a twisted nematic liquid crystal. In one sense a nematic system may be regarded as a twisted nematic system with infinite pitch. This implies that there is a continuously graded series of pitches. The pitch is the distance required along the axis of twist for a helicoidal liquid crystal to accomplish one complete rotation of its constituent molecules through 360° (or 2π radians).

4.1c Blue-phase liquid crystals

Blue-phase liquid crystals are a third type which is relevant to biological composites. Their structure is complicated but is related to that of chiral nematics. They differ, however, in that the twist in chiral nematics is planar, whereas that in blue-phases is cylindrical. Each cylinder may be imagined as a rolled-up planar helicoid. A model is given by Meiboom et al. (1981). The cylinders are arranged in three directions at right angles, as shown in Figure 4.1B, redrawn from Lepescheux (1988). Lepescheux compares the collagen architecture of the cuticle of a deep-sea polychaete annelid (*Paralvinella grasslei*) with a blue-phase liquid crystal. The cuticle, however, has only two sets of cylinders, the third direction being occupied by cylindrical microvillar extensions from the underlying epidermal cells. A three-dimensional system (Fig. 4.1A) which partly resembles a blue-phase is seen in the collagen bundles of fish scales (Giraud et al., 1978). However, the cylinders do not have coaxial helicoidal ultra-structure.

4.2 Relevant properties established from synthetic materials

4.2a Molecular asymmetry

We owe a debt of gratitude to chemists for establishing the basic principles of liquid crystals using synthetic compounds. One key feature of materials which form twisted nematic liquid crystals is their possession of *asymmetric* carbon atoms. These each attach to

126 / *Biomimicry: making liquid crystalline models*

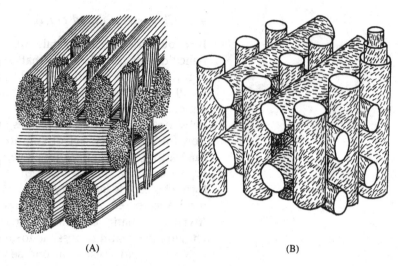

Figure 4.1. (A) A three-dimensional diagram of a small sample of collagen bundles in fish scales (twisted orthogonal architecture). Three major layers are arranged with a slight angular shift between the outer two and the middle one. A third system (with smaller-diameter fibres) traverses the major layers perpendicularly. Compare with part B. Redrawn from Giraud et al. (1978). (B) Diagram of a self-assembling blue-phase liquid crystal. Note the basic similarity to part A. Redrawn from Lepescheux (1988).

four different chemical groups, so that they may be arranged in two mirror-imaged forms (Fig. 3.22). This gives rise not only to laeval (L) and dextral (D) optical activity (described in Section 3.3b) but also to a chiral force which determines the direction of twist in rope-like fibrils. These, in turn, determine the sense of twist of chiral nematic liquid crystals which may self-assemble from them.

An excellent example is the synthetic polypeptide polybenzyl-glutamate (Fig. 3.23), which exists in either the D form (PBDG) or the L form (PBLG). These make fibrils with left- or right-handed helical twist, which can be observed in the electron microscope (Tachibana & Kambara, 1967). The fibrils have a diameter of 0.1–1.0 μm. In racemic mixtures the equal amounts of L- and D-isomers cancel each other out, resulting in fibrils with no twist. As in a manufactured rope, the sense of twist changes at each successive level in the hierarchy. Thus, PBLG forms D α-helices which themselves form superhelical fibrils with L-handed twist. The reason for this hierarchical change is that the molecules pack more tightly together when the twist alternates. Both PBLG and PBDG form twisted nematic liquid crystals, which have opposite senses of twist (Robinson, 1961).

A nematic liquid crystal of MBBA (methoxybenzylidene butylaniline), which is used in liquid crystal displays, may be caused to change into a twisted nematic liquid crystal by the addition of only very small amounts of an optically active substance such as cholesteryl benzoate (Bouligand & Livolant, 1984). The formulae of these two substances are given in Figure 4.2.

4.2b Energy aspects

That very low concentrations of optically active (asymmetrical) substances can induce twist in a nematic liquid crystal implies

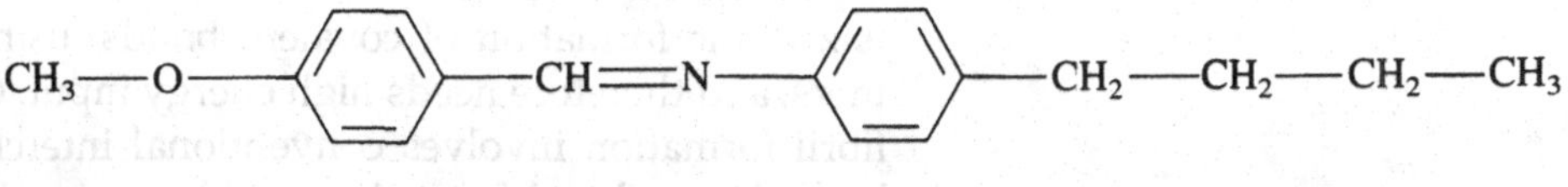

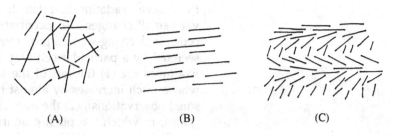

Figure 4.2. Synthetic liquid crystals. (A) Chemical formula of MBBA. This spontaneously forms a nematic liquid crystalline phase, with long straight molecules oriented mutually in parallel. We may think of it as a synthetic model of parallel-fibre systems (e.g. tendons). (B) Chemical formula of cholesteryl benzoate. If a small amount of cholesteryl benzoate is added to a nematic liquid crystal of MBBA, the latter becomes reoriented into a cholesteric liquid crystal with helicoidal texture. This may be regarded as a synthetic model of helicoidal systems (see Chapter 2). This illustrates why cholesteric liquid crystals are also known as twisted nematics.

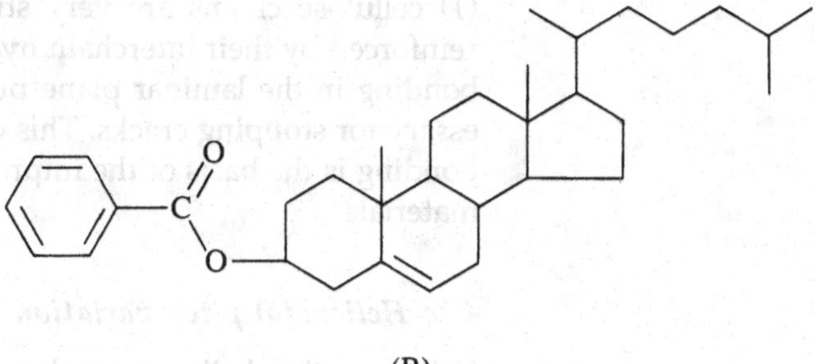

Figure 4.3. Energy of helicoid assembly. The energy required to convert a random arrangement of long molecules (A) into a mutually parallel nematic liquid crystal (B) is 100,000 times greater than the additional energy necessary to twist the nematic texture (B) into a helicoid (C). With so little extra energy required to construct helicoids, it is hardly surprising that they are so abundant in biology.

that only a small amount of energy is required for the formation of twisted nematic systems. The energy required to convert a nematic liquid crystal into a helicoidal liquid crystal with a pitch of 1.0 μm is 100,000 times less than the energy needed to induce parallel (nematic) order in an initially random system (Fig. 4.3).

Factors affecting the self-assembly of twisted nematic liquid crystals are Brownian motion, asymmetric carbon atoms, and the attempt to achieve the lowest amount of free energy. With a choice of right- or left-handed twists, the one with the lowest free energy will be selected. Each helicoidal composite has its own sense of twist, which does not come about by chance; it has a molecular basis. Further details on *absolute* sense of twist are given in Section 5.1e. Self-assembly systems in biological materials are more economical in energy requirements than are non-self-ordering structures, which involve enzymatic control and the hydrolysis of energy-rich phosphate bonds in ATP.

We may take plant cell walls as an example of levels of energy input during their formation: (1) Cellulose synthesis involves enzymatic formation of covalent bonds, using activated monomers, and therefore needs high energy input. (2) Cellulose microfibril formation involves conventional interchain crystallization by hydrogen bond formation, and hence moderate energy input. (3) Cholesteric liquid crystals are helicoids which self-assemble from less-ordered states. If they are involved in plant cell wall formation, they will require only very low additional energy input.

These various levels of developmental energy will relate directly to the mechanical properties of the composite formed. Thus, (1) cellulose chains are very strong in tension, and (2) this is reinforced by their interchain hydrogen bonding. (3) The weaker bonding in the laminar plane provides the weak interfaces necessary for stopping cracks. This combination of strong and weak bonding is the basis of the impressive performance of composite materials.

4.2c Helicoidal pitch variation

Cells secreting helicoidal architecture may vary its pitch. In Section 4.2b we saw that only small amounts of energy are needed to achieve gradation in pitch. It might therefore be caused by very small changes in concentration of an optically active molecule; such changes could prove very difficult to detect. A helicoid secreted by a particular cell may have sequentially larger pitch. Examples are (1) the test of the sea squirt *Halocynthia papillosa*, whose pitch increases by almost two orders of magnitude (personal observations), (2) the exocuticle of a silver beetle, *Plusiotis optima*, in which the pitch gradation causes reflection of wavelengths throughout the visible spectrum (Neville, 1977), (3) exocuticle of the locust *Schistocerca gregaria*, together with experimentally produced endocuticle which has helicoidal structure throughout its thickness (see Section 5.2h), and (4) cuticle of a spider *Cupiennius* (Barth, 1973).

By contrast, the pitch of the helicoidal cuticle secreted by the tobacco hornworm (*Manduca sexta*) decreases abruptly after the third day following the moult to the last larval instar (Wolfgang & Riddiford, 1986). Values for half pitch (180° rotation) change from 0.1–0.3 μm to 1.0–2.0 μm. This ten-fold change is associated with cessation in secretion of ten of the cuticular proteins, and an increase in quantity of the protein which they call LCP 27.

In the crab *Carcinus maenas*, the pitch of successively secreted cuticular layers varies from narrow to wide, and then back again to narrow (Bouligand, 1971). In other systems the pitch of secreted material may be constant, as, for example, in the physically coloured beetle *Lomaptera jamesi* (Neville & Caveney, 1969). In this beetle, helicoidal pitch determines colour, and there is a bilaterally symmetrical coloured pattern. Changes in colour from

Figure 4.4. Diagrams of two types of fault patterns (disclination defects) found exclusively in cholesteric liquid crystals and their solid-state analogues. These faults occur in helicoidal systems and are drawn here with lines representing each 180° rotation (= a lamella), and as seen between crossed polarizers. (A) Diagram of a triple-point fault. There is a rotation of lamellar direction due to insertion of one extra lamella (180°). (B) Diagram of a cofocal parabolic fault. There is a rotation due to deletion of a single lamella (180°). The rotation rectifies the disturbance of the even spacing by the decrease in amount of material. (C) Triple points and cofocal parabolae are usually found in pairs; each pair is designated as a translation dislocation. They are illustrated here as seen in beetle cuticle, appearing as 'fingerprint patterns' of chitin fibril orientations in layers stripped from the endocuticle of Dynastes cerebrosis. Large ducts (shown by dots) are the cause of these defects. A pair of translation dislocations is labelled (p = point of cofocal parabolae; t = triple point). The numbers relate to the layer sequences. Parts A and B are redrawn from Mazur et al. (1982). Part C is from Zelazny and Neville (1972), reprinted with permission of Pergamon Press, Ltd.

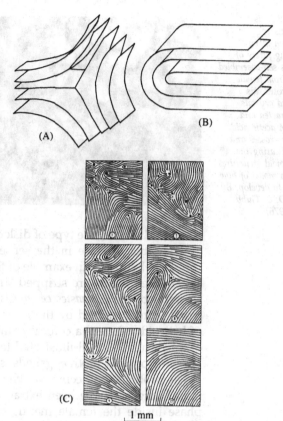

region to region follow faithfully the sequence of colours in the rainbow. In a mutant, the pattern is preserved, but the colours are reversed, so that pitch values for red become blue, and vice versa. Helicoidal pitch obeys bilateral symmetry, yet its sense of rotation does not; the helicoid is left-handed on both sides of the body.

Gradation in helicoidal pitch is also seen in the mucilage of the seed epidermis of quince, *Cydonia oblonga*, as seen in Figure 4.16. A possible mechanism for the origin of pitch gradation in a cholesteric liquid crystalline system is outlined in Section 6.2e.

4.2d Faults

Faults are found in natural fibrous composites. A helicoidal system which shows a rich variety of faults is the eggshell of the silk moth (Smith et al., 1971). Two of the types of faults occur in combination; these are the triple point and cofocal parabolae (Fig. 4.4A,B), which are together known as a translation dislocation. Such faults are also found in crustacean cuticle (Bouligand, 1969) and have been fully described and illustrated by Bouligand and

Figure 4.5. Building helicoids by self-assembly using mantis eggcase proteins. Light micrograph taken between crossed polarizing filters of cholesteric liquid crystals self-assembled from proteins extracted from female praying mantis (Sphodromantis tenuidentata). The liquid crystals are made by putting extracted and dried proteins into 0.1-M acetic acid. Spherulites with Maltese crosses and lines of birefringence indicating the half-pitch (180°) of helicoidal structures are seen. The system is a model of how insect cuticle is thought to develop. By kind permission of Dr. D. C. Gubb, from his Ph.D. thesis (1976).

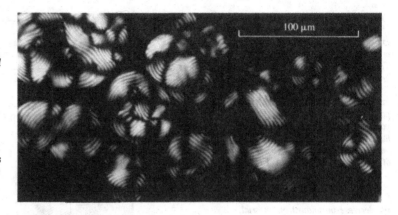

Kléman (1970). In one type of dislocation the pairs of triple points and parabolae rotate in the sense opposite to that of the surrounding helicoid. An example of this is seen (Zelazny & Neville, 1972) when layers are stripped successively from the cuticle of an adult beetle (*Dynastes cerebrosis*), as shown is Figure 4.4C. A triple point is caused by the insertion of an extra 180° half-turn of helicoid, whereas a cofocal parabola is caused by the deletion of a half-turn. A well-illustrated table of the types of faults now known from silk moth eggshells is given by Mazur et al. (1982).

The proteins (oothecins) which are destined to form a praying mantis eggcase have been extracted from the liquid crystalline phase inside the female mantis oothecal glands (Gubb, 1976). They have been dried to form a white powder which, when added to 0.1-M acetic acid, self-assembles into a helicoidal liquid crystal (Fig. 4.5). This shows faults, as expected. Both triple point and cofocal parabolae have been photographed (Gubb, 1976). Also, a radial fault like that seen previously in PBLG liquid crystals by Robinson (1961), has been seen in spherulites of self-assembled mantis oothecal proteins.

A helicoidal *spherulite* has the same twisted architecture as a planar helicoidal liquid crystal, except that it is spherical. Instead of being flat, the layers form a series of concentric spherical shells, like the skins of an onion. Unlike an onion, however, the fibres of each shell do not simply run from pole to pole. Rather, they twist successively through a small angle – always with the same sense of twist. A diagram is essential here; the one presented in Figure 4.6 is based on a drawing by Bouligand and Livolant (1984). With the exception of a single unique radius, each radius (in three dimensions) may be considered as an axis about which successive planes of parallel molecules rotate. It is a geometrical requirement that a helicoidal spherulite must possess a single radial fault.

In a now classical paper, Professor Sir Charles Frank applied the theory of elasticity to liquid crystals and was able math-

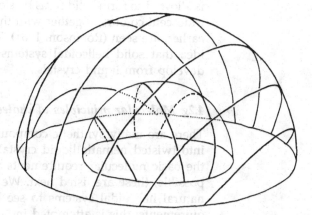

Figure 4.6. A highly simplified 3-D diagram of half a cholesteric spherulite. An intact spherulite would consist of many spherical shells of molecular chains or fibrils, each shell fitting over the previous (smaller) one like Russian dolls. Only two hemispherical shells are shown. The hoops represent the fibril direction in each shell, which rotates progressively – in this case anti-clockwise in layers situated farther from the observer. Redrawn from Bouligand and Livolant (1984).

ematically to predict the types of faults which must exist within them (Frank, 1958). His predictions have proven correct. They hold true equally for synthetic and naturally occurring liquid crystals. The significance of faults in biological systems is that the types which occur are found only in liquid crystals and their solidified equivalents; they do not occur in natural solid crystals (minerals). This provides indirect evidence in support of the idea that composite architecture in biological systems develops via a liquid crystalline stage.

4.2e Volume change in self-assembly

As noted by Bouligand and Giraud-Guille (1985), the phase change from random isotropic to liquid crystalline involves a decrease in volume. This applies irrespective of whether it is nematic, twisted nematic, or blue-phase. As a consequence of this, I would expect to see evidence in electron micrographs of the deposition zone in insect cuticles and in the periplasm of plant cell walls. If this is where self-assembly occurs, then these zones should be thicker than the most recently completed layer; this is seen to be so. A good example is seen in a micrograph of a section through the alga *Pithophora oedogonia* (Pearlmütter & Lembi, 1978).

4.2f Unusual optical properties

Several interesting optical properties are shared between twisted nematic liquid crystals and the exoskeletons of some adult scarab beetles. These are described in Section 3.3b. Provided that they have pitch values appropriate to interfere with visible light wavelengths, cholesteric liquid crystals show iridescent colours, reflection of circularly polarized light, and very high values of form optical rotation which disperses with wavelength. If PBLG is allowed to form solid films by slow evaporation of the solvent, iridescent colours, together with the associated properties noted earlier, are seen (Robinson, 1961). These similarities support the idea that solid helicoidal systems in plants and animals may develop from liquid crystals.

4.2g Molecular principles of cholesteric liquid crystals

There are many synthetic compounds which will self-assemble into twisted nematic liquid crystals. From them we may learn the basic molecular requirements which make such assemblies possible; these are listed next. We may then go on to consider natural helicoidal systems to see how far they meet these requirements; this is attempted in Section 5.1.

(1) One requirement, noted at the beginning of Section 4.2a, is *molecular asymmetry* in the form of an optically active carbon atom. The chiral nematics manufactured by British Drug Houses (BDH) serve as a synthetic example. Here biphenyl compounds are given chiral nematic properties by the addition of a terminal alkyl chain containing an asymmetric carbon (Fig. 4.7D). A natural example is C-5 in the sugar rings of cellulose and hemicellulose. Evidence for the importance of asymmetry is shown by the L and D forms of polybenzylglutamate, which form helicoids of opposite sense in liquid crystalline form (Robinson, 1966).

(2) Some chiral nematic liquid crystals are built with helical molecules, such as α-helical PBLG (Fig. 4.7A) and PBDG, and the double helical DNA (Robinson, 1966). Helical structure is not, however, an absolute requirement, because twisted nematic liquid crystals may be formed by non-helical molecules such as cholesteryl benzoate (Fig. 4.2B) and hydroxypropyl and acetoxypropyl celluloses (Fig. 4.7B,C).

(3) Long rod-like molecular shape (high axial ratio) does appear to be a common denominator. It is relevant here to note that isolated molecules of hemicellulose (glucuronoxylan) are seen to be straight and rod-like in the electron microscope (Vian et al., 1986).

(4) The development of a series of synthetic cellulose derivatives which are capable of self-assembly into twisted nematic liquid crystals with helicoidal structure (Werbowyj & Gray, 1976; Gray, 1983) is both invaluable and timely. They are made by reacting propylene oxide with cellulose and include hydroxypropyl cellulose and acetoxypropyl cellulose (Fig. 4.7B,C). I have found this work of great help when trying to understand the molecular basis of helicoidal structure in naturally occurring chiral nematic polysaccharides such as hemicelluloses. This is pursued in Section 5.1a.

The synthetic cellulose derivatives show the necessity for a *stiff straight molecular backbone*, so as to keep it parallel with

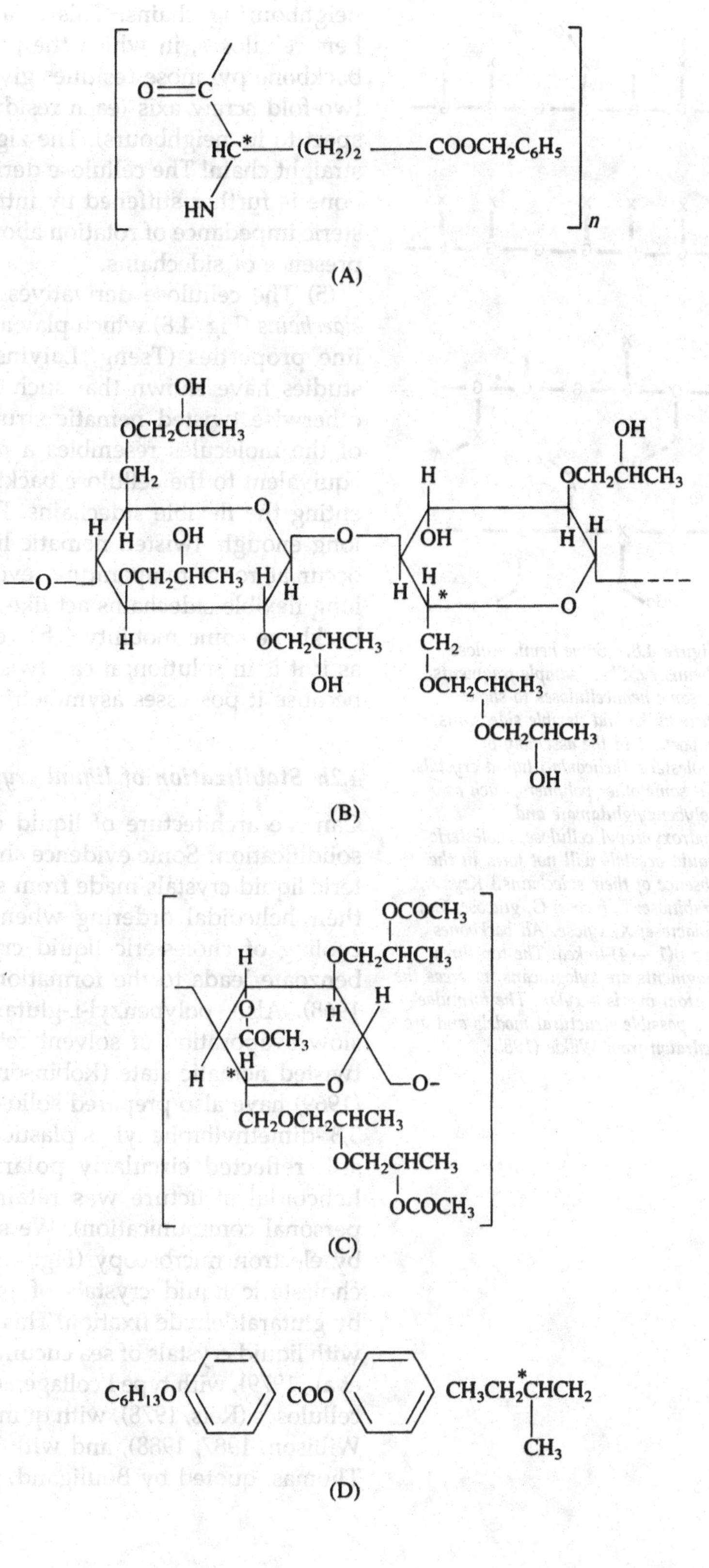

Figure 4.7. Some synthetic liquid crystalline substances. Asymmetrical carbon atoms are indicated by an asterisk. (A) In water, polybenzyl-(L or D)-glutamate forms cholesteric liquid crystals with a half-pitch (180°) appearing as lines resolvable in a light microscope. The benzyl sidechain is necessary for helicoid formation. Redrawn from Robinson (1966). (B) Hydroxypropyl cellulose, a cellulose derivative which forms cholesteric liquid crystals in water. The hydroxypropyl sidechain is necessary for helicoid formation. Redrawn from Werbowyj and Gray (1976). (C) Acetoxypropyl cellulose, another cellulose derivative which forms cholesteric liquid crystals when added to water. Redrawn from Gray (1983). (D) A chiral nematic compound redrawn from a brochure by BDH, whose chemists synthesized it. It is a biphenyl compound, made chiral by the asymmetrical carbon in the terminal alkyl chain.

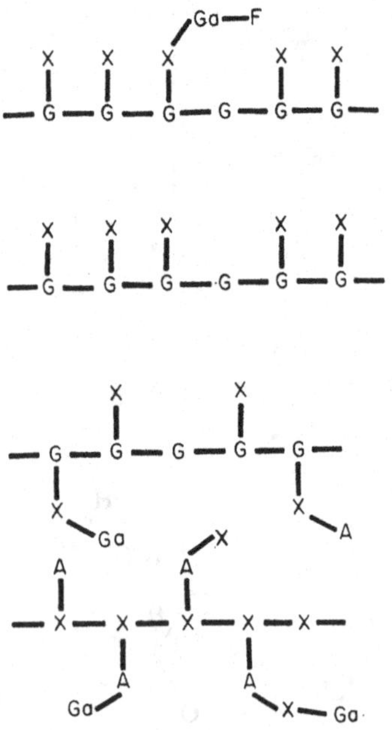

Figure 4.8. Some hemicellulose chemistry. Short sample fragments of some hemicelluloses to show their bulky and flexible sidechains, important in the assembly of cholesteric (helicoidal) liquid crystals. (In some other polymers, such as polybenzylglutamate and hydroxypropyl cellulose, cholesteric liquid crystals will not form in the absence of their sidechains.) Key: A, arabinose; F, fucose; G, glucose; Ga, galactose; X, xylose. All backbones are β(1 → 4)-linked. The top three fragments are xyloglucans, whereas the bottom one is a xylan. The formulae are possible structural models and are redrawn from Wilkie (1985).

neighbouring chains. This is also illustrated by cellulose and hemicelluloses, in which the β(1 → 4) glycosidic links between backbone pyranose residues give a zig-zag conformation with a two-fold screw axis (each residue rotates through 180° with respect to its neighbours). The zigs cancel the zags, resulting in a straight chain! The cellulose derivatives also show that the backbone is further stiffened by intrachain hydrogen bonds and by steric impedance of rotation about the glycosidic links due to the presence of sidechains.

(5) The cellulose derivatives have bulky, flexible and *mobile sidechains* (Fig. 4.8) which play a vital part in their liquid crystalline properties (Tseng, Laivins, & Gray, 1982). Comparative studies have shown that such sidechains must be present, for otherwise twisted nematic structure will not form. The shape of the molecules resembles a *pipe-cleaner*, with the wire being equivalent to the cellulose backbone chain, and the fuzz representing the flexible sidechains. Provided that the sidechains are long enough, twisted nematic liquid crystalline properties may occur at room temperature, even with no solvent present. The long flexible sidechains act like a solvent, allowing the cellulose backbone some mobility. The cellulose backbone chain behaves as if it is in solution; it can twist with respect to its neighbours, because it possesses asymmetric carbons.

4.2h Stabilization of liquid crystals

Can the architecture of liquid crystals be made permanent by solidification? Some evidence shows that it can, although cholesteric liquid crystals made from synthetic substances usually lose their helicoidal ordering when they solidify. However, rapid cooling of cholesteric liquid crystals of 2,4-dichlorocholesteryl benzoate leads to the formation of cholesteric glass (Fergason, 1968). Also, polybenzyl-L-glutamate allowed to form films by slow evaporation of solvent retains iridescent colours as in its twisted nematic state (Robinson, 1961). Samulski and Tobolsky (1969) have also prepared solid-state samples of PBLG by using 3,3'-dimethylbiphenyl as plasticizer. Some of these preparations still reflected circularly polarized light, indicating that the helicoidal structure was retained (Professor E. T. Samulski, personal communication). We showed (Neville & Luke, 1971b) by electron microscopy (Fig. 4.9) that it is possible to stabilize cholesteric liquid crystals of praying mantis eggcase proteins by glutaraldehyde fixation. This has subsequently been achieved with liquid crystals of sea cucumber defensive secretion (Dlugosz et al., 1979), with type I collagen (Giraud-Guille, 1989), with hemicelluloses (Reis, 1978), with quince seed mucilage (Abeysekera & Willison, 1987, 1988), and with blue-green algal mucilage (J. C. Thomas, quoted by Bouligand, 1978a).

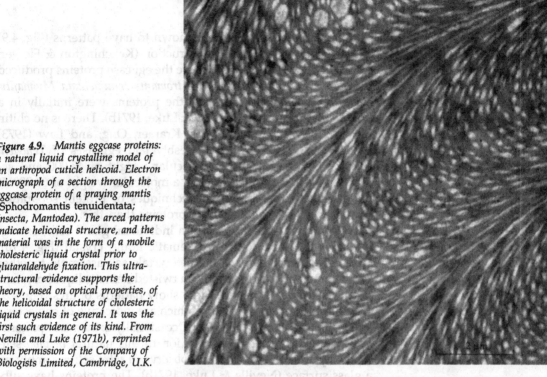

Figure 4.9. Mantis eggcase proteins: a natural liquid crystalline model of an arthropod cuticle helicoid. Electron micrograph of a section through the eggcase protein of a praying mantis (Sphodromantis tenuidentata; Insecta, Mantodea). The arced patterns indicate helicoidal structure, and the material was in the form of a mobile cholesteric liquid crystal prior to glutaraldehyde fixation. This ultrastructural evidence supports the theory, based on optical properties, of the helicoidal structure of cholesteric liquid crystals in general. It was the first such evidence of its kind. From Neville and Luke (1971b), reprinted with permission of the Company of Biologists Limited, Cambridge, U.K.

4.3 Natural liquid crystal models

We are fortunate to have access to a number of examples of naturally occurring liquid crystals of several types (helicoidal, nematic, orthogonal). The available information is assembled together in this section. The existence of the models makes feasible the underlying theme in this book: that a self-assembling liquid crystalline phase precedes formation of solid materials with the same types of architecture. The liquid crystalline models can also be used to explore potential mechanisms, such as concentration, temperature, and pH, by which neighbouring cells may exert control over extracellular materials.

Specifically, there are models for some of the main skeletal systems of biology. Insect cuticle may be mimicked by praying mantis eggcase protein; bone and vertebrate cornea may be mimicked by sea cucumber defensive proteins; and plant cell walls may be mimicked by quince mucilage. The sources of these liquid crystals are described in Chapter 2. They are grouped here according to whether the fibrous component is protein or polysaccharide. An attempt was made to bring the twisted nematic nature of some proteins to the notice of liquid crystal chemists (Neville, 1981). Since then our knowledge of collagen and moth eggshell proteins has increased considerably, and a promising start has been made on liquid crystalline polysaccharides.

4.3a Mantis oothecal proteins

The first proteins which were shown to have patterns (Fig. 4.9) characteristic of helicoidal construction (Kenchington & Flower, 1969; Neville & Luke, 1971b) were the eggcase proteins produced by female praying mantids (*Sphodromantis tenuidentata*, *Miomantis monarcha*). We recognized that the proteins were initially in a liquid crystalline state (Neville & Luke, 1971b). There is no chitin associated with these proteins. Kramer, Ong, and Law (1973) isolated three proteins from freshly laid eggcases of *Tenodera sinensis*. Oothecin I had a molecular weight of 43,000 and accounted for 80%; oothecin II had a molecular weight of 60,000. It is very likely that improved techniques of protein separation would now reveal many more proteins than this.

The proteins are secreted from individual gland cells, emerging initially in the form of a nematic liquid crystal (Fig. 4.10). These individual nematic liquid crystals self-assemble together in the lumen of the gland to form twisted nematic liquid crystals; when fixed and sectioned, these show patterns of arcs (Fig. 4.9) with double spirals (Fig. 4.11) which are indicative of cholesteric spherulitic structure (Fig. 4.12). Occasionally orthogonal patterns are also seen (Fig. 2.4). In addition to this natural *in vivo* self-assembly we were also able to observe *in vitro* self-assembly on a glass surface (Neville & Luke, 1971b). The proteins have subsequently been extracted and freeze-dried in amyl acetate with solid carbon dioxide, and the water sublimed off under vacuum. When water was replaced, the proteins gradually self-assembled back into the twisted nematic liquid crystalline state (Gubb, 1976). Ionic interactions do not seem to be important for self-assembly, because the twisted nematic structure survives in 0.2-M sodium sulphate solution. Neither calcium ions nor its chelator EGTA affect the self-assembly. But pH does affect it, causing the twisted nematic phase to become optically isotropic below pH 5.0.

Water is required for the liquid crystalline state to form. Calculations based on amino acid composition suggest that these proteins have a very high hydrophilic index (Neville, 1981). Application of 200 atmospheres pressure increases the number of water molecules absorbed onto the ionic amino acids, causing the water to occupy a lower volume than when in free solution. The liquid crystalline phase becomes isotropic (Gubb, 1976).

Consideration of molecular shape based on X-ray diffraction suggests a two-stranded coiled coil of α-helices (Rudall, 1956). Polar residues tend to occupy the surface of a protein, so that a high polar content will lead to a high axial ratio. A long coiled coil shape would be appropriate for forming a twisted nematic

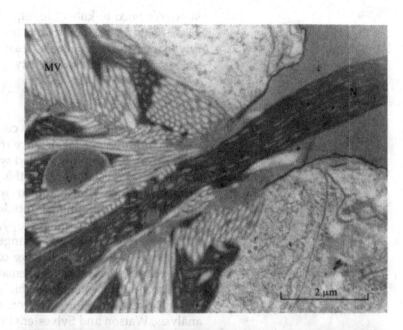

Figure 4.10. Electron micrograph of a section through a gland cell which is actively secreting eggcase protein in a female praying mantis (Miomantis monarcha; Insecta, Mantodea). The protein is secreted between the microvilli (MV) and emerges as a nematic liquid crystal (N) with molecules in parallel. Only when it reaches the lumen of the gland (not visible here) does it self-assemble into a cholesteric liquid crystal, with an arced pattern as in Figure 4.9. This is a sort of 'action' electron micrograph. From Neville and Luke (1971b), reprinted with permission of the Company of Biologists Limited, Cambridge, U.K.

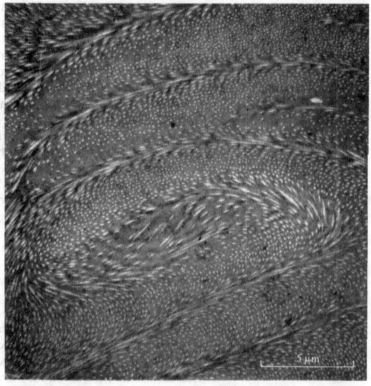

Figure 4.11. As for Figure 4.9, but for a section passing through the centre of a liquid crystalline spherulite. The double spiral pattern is characteristic of a helicoidal spherulite. From Neville and Luke (1971b), reprinted with permission of the Company of Biologists Limited, Cambridge, U.K.

Figure 4.12. A diagram of the pattern expected in an equatorial section through a cholesteric liquid crystal spherulite. This type of pattern is seen in sections of curved helicoids (e.g. crab cuticular tubercles, insect eye lenses, mantis oothecal spherulites). Compare with Figure 4.10. Redrawn from Bouligand (1965).

state. We need to know the sequences of these proteins, so as to relate them to secondary structure and thence to liquid crystalline assembly. They have at least the merit of being readily extractable in reasonable quantity.

4.3b Sea cucumber collagen

The sticky secretion of a sea cucumber (*Holothuria forskali*) is ejected from the gut for defence from predators or for food capture. Before ejection it is located within the Cuvierian tubules. It is in a liquid crystalline state; this can be shown by the changes in birefringence which appear when the tubes are distorted. Fixation and sectioning of tubules and their contents *in situ*, followed by electron microscopy, shows that the secretion is initially in twisted nematic arrangement (Dlugosz et al., 1979). It also shows regions which have orthogonal structure (Fig. 2.5). Upon release, it changes to a nematic (uniaxial and parallel) fibre and also undergoes considerable extension.

By X-ray diffraction, electron microscopy, and amino acid analysis, Watson and Sylvester (1956) have identified the fibrous component as collagen. The banding pattern and 68-nm axial repeat are closely similar to those of type I collagen, such as that in rat tail tendon (Bailey et al., 1982). These collagen fibrils have a mean diameter of about 60 nm and a very irregular outline in cross-section. They are accompanied by about forty times more glycosaminoglycan than in rat tail collagen; most of this is heparan sulphate, together with some chondroitin sulphate. Bailey et al. (1982) suggest that the proteoglycans form a sheath around the collagen fibrils, preventing interfibrillar cross-linking. This would account for the great extensibility of the collagen by shearing of fibrils within fibres. Perhaps the carbohydrates are also involved in the twisted nematic assembly? This system represents the first recorded natural example of twisted nematic liquid crystalline collagen.

4.3c Vertebrate collagen

Collagen is a protein with a very long narrow molecule (300 nm × 1.5 nm). Three separate molecular chains are twisted to form a triple helix (Fig. 4.13) which behaves as a semi-rigid rod. The collagen helix is not to be confused (as it sometimes is in the literature) either with an α-helix or with coiled α-helices.

Collagen represents a classic example in self-assembly – at least as far as the fibril level. However, despite many attempts, it is only recently that convincing self-assembly of liquid crystalline collagen has been achieved: by Bouligand et al. (1985) for orthogonal and for limited twisted nematic examples, and by Giraud-Guille (1989) for extensive and convincing examples of

Figure 4.13. (A) End-on view of a collagen triple helix. Each collagen molecular chain coils to form a left-handed helix. Three of these helices then pack together like the components of rope to form a right-handed triple helix. They pack very tightly because of the size and regular occurrence of glycine (the smallest possible amino acid), which forms every third residue. Each left-handed helix has three amino acids per turn. The tight packing is enabled by the positioning of the glycines in the centre of the rope. (B) Collagen triple helix (a coiled coil) with an individual component helix drawn in. This triple helical rope has grooves which may well dictate the packing of layers of such ropes to form a helicoid (see Fig. 6.7). From Woodhead-Galloway (1980), reproduced by permission of Edward Arnold (Publishers) Limited, and of the author.

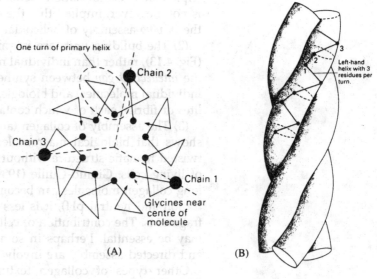

twisted nematic structure. In an earlier study she had achieved an *in vitro* assembly which was almost helicoidal (Giraud-Guille, 1987). Her trick in making cholesteric collagen lay in using fragments of suitable length. Calculations by Flory (1961) had shown that alignment of rigid rod-shaped molecules in concentrated solution requires an axial ratio (length/diameter) of about 10. Collagen from calf skin has an axial ratio which is far too high for self-assembly (300 nm/1.5 nm = 200). Giraud-Guille therefore disolved collagen in 0.5-M acetic acid and fragmented it down by sonication at 20 kHz to a mean length of 18 nm. These short fragments of triple helix then have sufficient freedom of movement (even in highly concentrated solution) to form a cholesteric liquid crystal – given a few hours assembly time. The preparations were concentrated by slow evaporation under vacuum. Polarizing light microscopy then revealed the characteristic banding patterns and faults of a cholesteric texture. It is seen to be liquid crystalline by its viscosity and mobility on distortion. The appearance is indistinguishable from that of self-assembled praying mantis protein described in Section 4.3a. Cofocal parabolae, triple points, spherulites, a half-pitch (layer spacing) of 1–4 μm, and some planar (flat) cholesteric domains are all in evidence. Vacuum dehydrated films examined in the electron microscope showed areas with fibre banding in perfect alignment – just like the proteins in the periostracum of whelk shells (Fig. 2.27).

I believe that this landmark of achievement by Giraud-Guille has some important implications:

(1) Collagen is competent, even in the absence of a glycosaminoglycan matrix, to form a cholesteric liquid crystal. The

triple helical conformation seems ideal for helicoid formation. It is not, however, implied that the carbohydrates play no part in the *in vivo* assembly of helicoidal collagen.

(2) The building units are fragments of collagen triple helices (Fig. 4.13), rather than individual molecules. This helps to bridge the *intellectual gap* between synthetic liquid crystals (built with individual molecules) and biological helicoids (built with crystallites or fibrils, each of which contains several molecular chains).

(3) The assembly of collagen (and of praying mantis proteins) shows that biological macromolecules are capable of forming twisted nematic structure without the need for the intervention of living cells. Giraud-Guille (1989) notes, however, that while long collagen molecules can become aligned (given low concentration and neutral pH), it is less easy for them than for small fragments. The contribution of cells to collagen fibril orientation may be essential. Perhaps in some systems both self-assembly and directed assembly are involved (see Sections 5.1 and 5.2)?

Other types of collagen texture have been built by self-assembly. There are reports of orthogonal orientation of collagen fibres in self-assembled systems (Trelstad, Hayashi, & Gross, 1976; Bouligand et al., 1985). These orthogonal textures do not, however, cover very extensive volumes. Orthogonal systems are shown better in fixed and sectioned samples from two non-vertebrate sources: sea cucumber collagen (Fig. 2.5) and praying mantis eggcase protein (Fig. 2.4). These are in a liquid crystalline state prior to fixation.

Various other collagen liquid crystalline textures have been made by self-assembly and superbly illustrated (Bouligand et al., 1985). They include cylindrical twists and toroidal twists. A section of a chitin/resilin system from a locust (Fig. 3.3) shows great similarity to a collagen cylindrical twist texture.

We now have fairly complete knowledge of the structural hierarchy of collagen; for a concise and informed review, see Woodhead-Galloway (1980). It can be extended to include the liquid crystalline textures to show where they fit into the hierarchy. The amino acid composition of collagen includes extensive glycine-proline-hydroxyproline sequences. This forbids formation of α-helices. A left-handed collagen helix is formed, three of which entwine to form a right-handed triple helix – the protofibril. Better molecular packing and hence greater tensile strength are achieved if the sense of twist is reversed at each structural level. The same principle is used in the construction of rope. Depending upon which model is adopted, either five or six protofibrils coil to form a microfibril. The protofibrils are staggered with respect to each other, and a gap is left between the end of one microfibril and the start of another. These gaps are the sites of nucleation for crystals of hydroxyapatite to be deposited in bone. The interaction of mineral and collagen creates the property of piezoelectricity at this level. The microfibrils pack to form domains with a tetragonal lattice, with each domain constituting a subfibril. Several subfibrils co-operate to form a fibril. Only at the fibrillar level does interaction between collagen and matrix occur. It is at this level also that fibrils become organized into parallel fibres, orthogonal trellises, helicoidal superply, and cylindrical or toroidal twists. Collagen is a huge topic, and further reading is recommended. For discussion of collagen orientation in the light of liquid crystals, see Bouligand and Giraud-Guille (1985); for bone collagen see Giraud-Guille (1987, 1989); and for a summary of collagen interfaces in invertebrates see Gaill (1990).

4.3d Moth eggshell proteins

Indirect evidence suggests that a liquid crystalline stage may precede formation of helicoidal structure in moth eggshells (Smith et al., 1971). Sections of eggshell in an early phase of development show protein microfibrils in less regular array (Fig. 4.14A) than at a later stage (Fig. 4.14B). The variety and abundance of disclination faults in this material are strongly indicative of a liquid crystalline phase. Very large amounts of time, money, and effort have gone into the extraction, separation, and sequencing of the numerous (nearly 200) proteins in moth eggshells, by a team at Harvard led by Professor Kafatos. An interpretation of their results in terms of trying to understand the molecular basis of helicoid formation is given in Section 5.1. As a co-author of the pilot paper which triggered this work, I confess to occasional feelings of anxiety. But if it works, it should yield the first complete molecular explanation of how a natural helicoid is formed. The time is ripe for attempts to self-assemble individual extracted eggshell proteins, and combinations of them, into liquid crystalline phases.

4.3e Hemicelluloses from plant cell walls

Those who believe that helicoidal plant cell walls may arise by self-assembly also think that the hemicelluloses in the matrix are the key architects. The evidence for this is presented in Section 5.1a. Yet the ability of hemicelluloses to self-assemble into cholesteric liquid crystals has seldom been demonstrated. Reis (1978) used alkaline extraction to obtain hemicelluloses from the cell walls of mung bean (*Phaseolus aureus*) hypocotyls. She succeeded in making the extract self-assemble into a gel with multilayered nodules. Electron microscopy of sections of the nodules showed arced patterning characteristic of helicoidal structure. Ionic and pH conditions are critical to this assembly. Sam Levy (personal communication) has similarly obtained arced patterns by precipitating in alcohol an alkali-extracted fraction

142 / *Biomimicry: making liquid crystalline models*

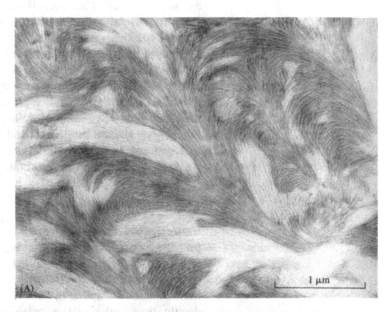

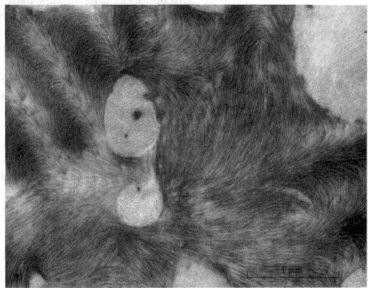

Figure 4.14. (A) *Electron micrograph of a section through the wall of an egg from a silk moth (*Hyalophora cecropia*; Insecta, Lepidoptera). This region shows protein microfibrils in an array which is far less regular than a helicoid. It is interpreted as a possible liquid crystalline stage which later develops into a helicoidal layering (as in part B). A very large amount of work in Harvard, Athens, and Crete followed our pilot paper (Smith et al., 1971). It recognizes the exciting possibility of understanding the protein conformations which give rise to the chitin-free helicoidal eggshells of butterflies and moths. By kind permission of Professor W. Telfer. (B) As for part A, but for a later stage of eggshell development, showing arced patterns characteristic of helicoidal structure. By kind permission of Professor W. Telfer.*

from the cell walls of a stonewort (*Nitella translucens*). The components which formed the self-assembly were 20–30 nm in length. This seems to meet the dimensional criterion of Flory (1961) summarized in Section 4.3c.

4.3f Quince mucilage: a paradigm model of plant cell wall?

Recently a system has been discovered which gives the first *direct* evidence in support of the self-assembly hypothesis for helicoidal plant cell walls. It occurs in the epidermal cells of seeds of the quince (*Cydonia oblonga*; see Section 2.4) and is in the form of a cholesteric liquid crystalline mucilage (Abeysekera & Willison, 1987). Electron microscope investigation of different stages of development shows that the mucilage first appears amorphous, then with randomly intermingled cellulose microfibrils, and finally with the microfibrils oriented with arced patterning typical of helicoids (Figs. 4.15 and 4.16). The arced patterning arises in a region of the periplasm well away from both the completed cell wall and the cell plasma membrane. This is evidence for self-assembly, rather than a layer-by-layer protoplast-directed assembly (Abeysekera & Willison, 1988).

When the periplasmic contents are placed in distilled water, they swell to form a mucilaginous gel. The contents clearly are not strongly bonded together. Taken together with the birefringent domains seen between crossed polarizers, this indicates that the material is in the liquid crystalline state. Furthermore, these observations may be made at any time after sixty days following flowering (Willison & Abeysekera, 1988). The system therefore *remains* in a liquid crystalline form. When dried on a grid, the material shows cellulose microfibrils in the electron microscope.

Normally, the periplasm of plant cells is a thin zone (about 0.5 μm thick) sandwiched between the outer surface of the cell plasma membrane and the inner surface of the most recently completed extracellular wall. In quince seed epidermis, however, the cytoplasm progressively shrinks (Willison & Abeysekera, 1989); as it does so, the mucilage fills the available space (Fig. 4.17). Abeysekera and Willison (1990) used serial sectioning, transmission electron microscopy, montage, and 3-D reconstruction to plot the shape and architecture of the secretion from individual cells. In most cases the periplasm contained a single helicoidal array, sometimes with a spiral pattern (Fig. 4.15), indicative of a cholesteric spherulite. Occasionally, however, a nematic liquid crystalline region occurred in the same periplasm as a helicoidal region. The significance of this is considered in Section 6.1.

Schnepf and Deichgräber (1983a,b) have reasoned that periplasmic mucilage, wherever it is found, is a specialized example of a plant secondary wall – provided that it contains cellulose microfibrils. The study of mucilage formation might therefore help in the understanding of cell wall development. Those of us who believe in helicoidal walls would agree: The quince mucilage material could provide important direct evidence on the chemistry and morphogenesis of helicoidal cell walls. Until now we have had to rely on circumstantial evidence. A promising start has been made on the chemistry of the matrix of this mucilage (Section 5.1a).

Recently, Reis et al. (1991) have disassembled cellulose and glucuronoxylan hemicellulose from quince mucilage and then

Figure 4.15. A remarkable natural liquid crystalline model of plant cell walls. Electron micrograph of a section through the mucilage produced by epidermal cells of the seeds of a quince bush (Cydonia oblonga). The double spiral pattern is typical of a section through the centre of a cholesteric liquid crystalline spherulite with helicoidal structure. It consists of cellulose microfibrils in a polysaccharide matrix and represents a very large periplasm (see Fig. 4.17). In normal cells, the periplasm is a very narrow zone between a plant cell plasma membrane and the cell wall outside it. It is believed that the quince mucilage represents a model of the mode of formation of a helicoidal cell wall; these are thought to self-assemble into a liquid crystal and then solidify (Neville et al., 1976). From Abeysekera and Willison (1987), reprinted by permission of Academic Press Limited, London, and of the authors.

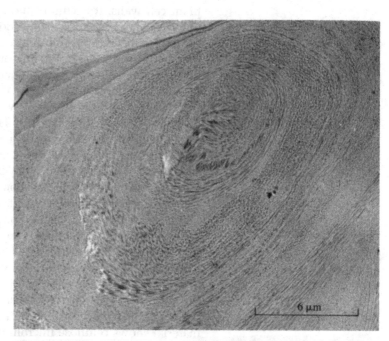

Figure 4.16. As for Figure 4.15, showing the ranks of arced patterns typical of helicoidal structure. The cellulose microfibrils are heavily stained. The system is cholesteric liquid crystalline and is an excellent example of gradation of pitch, of angular rotation, and of number of planes of microfibrils making up an arc. From Abeysekera and Willison (1990), by permission of Editions Scientifiques Elsevier, Paris, and of the authors.

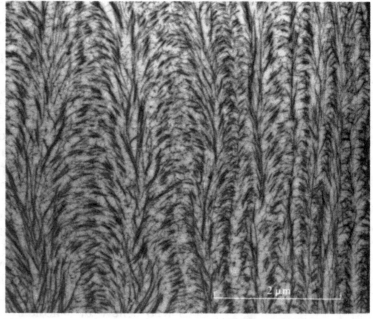

reassembled them into a helicoidal system. This is the first time that *microfibrils* of cellulose have been self-assembled with hemicellulose to form helicoids. The aqueous extract was reconcentrated by centrifugation and dehydration, and the helicoids were detected by both polarization and electron microscopy. Probes were used *in situ* to show the close association of glucuron-

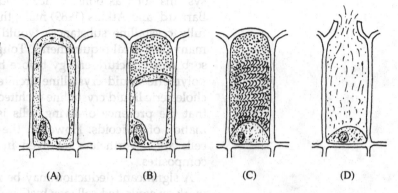

Figure 4.17. Diagrams of stages in development of a seed epidermal cell in quince fruits (Cydonia oblonga). (A) A fully extended cell with large central vacuole. (B) The cell cytoplasm and its central vacuole are shrinking to make way for a large periplasm (dotted), initially filled with amorphous materials. (C) The periplasm has enlarged further, and now shows ranks of arced patterns indicative of helicoidal structure. The periplasm is in the form of a cholesteric liquid crystal, with cellulose microfibrils creating the patterns. The texture moulds to the shape of the containing cell walls. (D) The seed epidermal outer cell wall has burst. The periplasm swells and oozes out of the cell as mucilage. Redrawn from Willison and Abeysekera (1989).

oxylans around each cellulose microfibril. Free carboxyl groups in the xylans were labelled with cationized colloidal gold, while cellulose was labelled with cellobiohydrolase-gold complex. In the TEM, both types of probes showed close alignment of gold particles along the microfibrils, indicating that the glucuronoxylan forms a sheath around each cellulose microfibril. Reis and her colleagues note that the quince mucilage system is particularly valuable because its components are accessible by aqueous extraction; the polysaccharide polymers therefore remain in a natural state.

4.3g Cyanobacterial mucilage

Quince seed mucilage is not the only type with helicoidal structure. Mucilage secreted by the blue-green algae *Rivularia atra* and *Chroococcus minutus* can be fixed for electron microscopy with glutaraldehyde. Very convincing micrographs of ranks of arced patterning, indicating helicoidal texture with a pitch of 0.4 μm, have been obtained by J. C. Thomas (quoted and illustrated by Bouligand, 1978a). Perhaps the cholesteric structure stabilizes such mucilage, as it does the foam of mantis eggcase. Does the secretion of helicoidal mucilage by these primitive blue-green algae reveal an evolutionary step towards the use of this texture for cell walls of higher forms? Priority should be given to chemical analysis of this convenient form of helicoidal material.

4.3h Conclusions

It has been previously noted (Neville, 1981) that 'any biological structure which is helicoidal is likely either to be cholesteric or to have passed through a cholesteric phase during its development'. Liquid crystalline architectures match those required for helicoidal, orthogonal, and parallel textures in living systems. There are even several natural liquid crystalline models for major systems such as bone, cuticle, and plant cell wall. Gathercole, Barnard, and Atkins (1989) make the interesting observation that fully crystalline substances would be too hard and brittle for many biological requirements. Tougher materials, which can absorb more fracture energy before breaking, may be made from polymeric liquid crystalline architectures. The self-assembly of cholesteric liquid crystalline architecture from dry proteins shows that the presence of living cells is not obligatory for the formation of helicoids. However, the next chapter will show that cells are nevertheless involved in the development of fibrous composites.

A significant deduction may be made from the self-assembly work on sonicated collagen by Giraud-Guille (1989). A cholesteric liquid crystalline assembly can be made with units of construction which are larger than individual molecular chains – in this case with fragments of triple helix. This overcomes one of the differences between synthetic liquid crystals (built from individual molecules) and biological fibrous composites (built of microfibrils in a matrix). It can also be noted that cholesteric liquid crystals have been assembled using whole rods of *Narcissus* mosaic virus (Wilson & Tollin, 1970). Also, the important recent work by Reis et al. (1991) shows that glucuronoxylans interact with cellulose in its microfibrillar crystalline state.

5
How is fibre orientation controlled?

It is convenient to distinguish between primary and secondary fibre orientations (Neville, 1967c). The first two sections of this chapter (5.1 and 5.2) concern primary orientation, in which fibres are laid down in their final positions. In Section 5.3, secondary orientation is considered; here fibres are laid down in one type of orientation, but afterwards are changed by some physiological force to a new position. There are two major categories of primary fibre orientation mechanisms. In Section 5.1, fibres are positioned by *self-assembly*, powered by intermolecular forces. In Section 5.2, fibres are positioned by *directed assembly*, driven by some type of cellular mechanism.

Where does the assembly of extracellular architecture take place? In the case of plant cell walls, cellulose is synthesized on the plasma membrane, whereas hemicellulose is synthesized in the Golgi apparatus and then transported by membrane flow to the plasma membrane. These and other components mix in the periplasm, which is a narrow region constrained between the most recently deposited cell wall layers (on its outer side) and the cell plasma membrane (on its inner side). Such mixing of matrix and microfibrils is confirmed by direct observations on the developing periplasm of quince seed epidermis (Section 4.3f). It is fortunate for future work that the quince periplasm is so large; a major experimental problem with most systems is that the zone where assembly takes place is so thin.

The existence of helicoidal structure in some plant cell walls is becoming gradually (if sometimes reluctantly) accepted. This requires that somewhere there is a rotary mechanism. If this is by self-assembly, then it is likely to be located in the periplasmic domain; but if assembly is directed, then it may occur in the cytoplasm. A third (but unlikely?) possibility is that there is some as yet unknown membrane-bound rotary motor, perhaps in the membrane rosettes where cellulose is polymerized.

The equivalent of the plant periplasm in insects is the cuticle deposition zone adjacent to the epidermis. It was suggested

(Neville, 1975a) that the deposition zone might have some degree of permanence, since the epidermal cells could hardly adhere to a liquid crystal. This suggestion has received support from Wolfgang, Fristrom, and Fristrom (1987), who showed that the deposition zone is immunologically distinct from completed cuticle. Traces of parallel or helicoidal structure are seen in the deposition zone, depending upon the time of day when fixation took place (Neville & Luke, 1969a,b).

The chorion of moth eggshells is secreted by the follicle cells of the female moth. Assembly of the helicoidal structure takes place at some distance from the zone of secretion (Smith et al., 1971). Furthermore, the proteins have to pass through a sieve layer positioned between the follicle cells and the chorion. There can be no cell-directed assembly in this instance.

Similarly, in developing earthworm cuticle, matrix filaments become orthogonally ordered in that part of the newly secreted cuticle which lies farthest from the underlying epidermal cells (Humphreys & Porter, 1976). Collagen fibrils then form on this scaffolding, once more starting farthest from the cells. This is evidence against the involvement of directed assembly in this particular orthogonal system.

In the case of developing bird cornea, each new layer of collagen fibrils is laid down in the primary stroma immediately to the inside of the layer of epithelial cells which covers the outside surface of the eye. This has been established by autoradiographic studies (Trelstad & Coulombre, 1971).

5.1 Self-assembly: hypotheses based on molecular shape

This section considers our knowledge of molecular structures in several fibrous composites, in the light of self-assembly of helicoids.

5.1a Plant cell walls: hemicelluloses

Which plant wall molecules are competent to create helicoidal structure by self-assembly? The candidates are cellulose, pectins, lignin, and extensin; they have been considered in a previous publication (Neville, 1988a). While *cellulose* has a straight, stiff backbone, it does not have bulky sidechains. The backbone chains are therefore able to approach each other closely. It is relevant to recognize the distinction here between homopolymers and heteropolymers. A homopolymer contains just one type of residue and can crystallize with the chains closely packed in parallel to form crystallites; cellulose is a homopolymer. By contrast, heteropolymers (such as hemicelluloses) contain more than one kind of residue and do not form crystallites. The neighbouring chains of cellulose crystallize by hydrogen bonding in planar sheets which are themselves held together by London dispersion

5.1 Self-assembly: hypotheses / 149

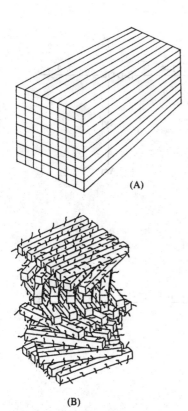

Figure 5.1. *Cellulose versus hemicellulose architecture. (A) Highly simplified diagram of a microfibril composed of cellulose molecules. Each cellulose chain is represented by a rod with a square cross-section. Cellulose molecules are able to pack in this parallel and highly compact way because they have only small side groups. (B) Highly simplified diagram of a group of hemicellulose molecules, for comparison with part A. Each hemicellulose chain is represented by a rod with a square cross-section and with flexible, bulky sidechains (shown as squiggles). The molecular backbones behave as if they lie in a solution of sidechains. Prior to crystallization, this allows the hemicellulose backbones some freedom of rotary movement (not present in cellulose). The successive layers of hemicellulose are able to move through a small angle, driven by twisting forces generated by asymmetry in the backbones. Instead of crystallizing in parallel as in cellulose, as in part A, crystallization in hemicelluloses is hindered, so that a 'hairy' helicoid is able to form. From Neville (1988a), by permission of Academic Press Limited, London.*

forces to form a crystallite (Gardner & Blackwell, 1974). There is no angular twist between the sheets (Fig. 5.1A), such as would be required by a helicoid-former. It has been noted that cellulose does not form liquid crystals except in unphysiological conditions (Willison & Abeysekera, 1989) or when mixed with hemicelluloses.

We may also eliminate *pectins*, whose molecular structure is inconsistent with the needs of regular helicoids. Pectins form an open-meshed gel (Jarvis, 1984), with junctions stabilized by calcium ions and connected by flexible chains. Water is included to form a soluble gel. The chains have $\alpha(1 \rightarrow 2)$ and $\alpha(1 \rightarrow 4)$ linkages with a three-fold screw axis (i.e. an angle of rotation of 120° between consecutive sugars). This prevents pectins from bonding strongly to cellulose, which has $\beta(1 \rightarrow 4)$ linkages giving a two-fold screw axis (180° between residues). Because cellulose cannot by itself form helicoids, any 'helper' molecules able to manipulate it into such structure would need to bond to cellulose. A further reason why pectins should not act as helicoid-generators may be deduced from two observations in the literature. Chemical analysis of cell walls from stalks of *Cyperus papyrus* reveals that they do not contain pectins (Buchala & Meier, 1972); they do, however, have cell walls with helicoidal structure (Mosiniak & Roland, 1986).

Lignin may be ruled out as a helicoid-generator using the same rationale; helicoidal structure is found in non-lignified walls such as those from *Chara* (Neville et al., 1976) and *Nitella* (Neville & Levy, 1984).

On the face of it, the glycoprotein *extensin* seems like a potential candidate, because it consists of a polyproline II helix backbone festooned with sidechains each containing four arabinose sugars. Extensin may, however, be eliminated for the same reason as that just given for lignin.

By elimination, *hemicelluloses* emerge as the favourite candidate for helicoid generation in plant walls (Neville, 1988a); there are several reasons:

(1) They are abundant, forming up to 40% of the weight of some plant walls.

(2) Hemicelluloses have a pipe-cleaner structure with a stiff backbone chain (Darvill et al., 1980; MacLachlan, 1985). Numerous flexible and bulky sidechains occur (Fig. 5.2). These are precisely the criteria for formation of cholesteric liquid crystals, as established on synthetic cellulose derivatives in Section 4.2g. The molecules contain optically active carbon atoms, and the sidechains contain xylose, galactose, fucose, and arabinose. The backbones may be composed either of glucose (as in xyloglucans) or of xylose (as in glucuronoxylans). The sidechain linkages are such as to permit mobility: $\alpha(1 \rightarrow 2)$, $\alpha(1 \rightarrow 3)$, $\alpha(1 \rightarrow 6)$. There are no $\beta(1 \rightarrow 4)$ linkages in the sidechains, which

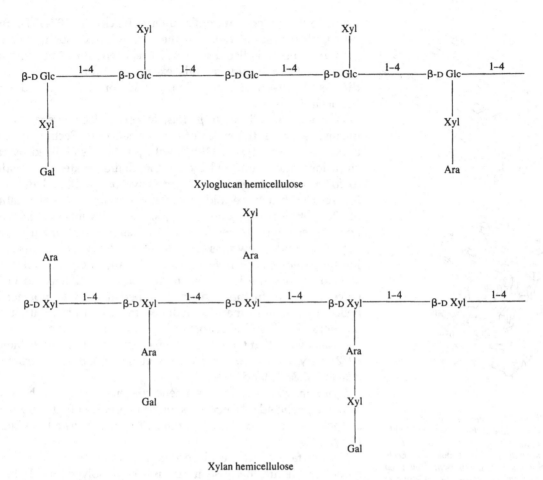

Figure 5.2. Hemicellulose chemistry. Parts of chains of two types of hemicelluloses found in plant cell walls. The xylan is a notional composite, to illustrate a typical selection of known sidechains. Glc, glucose; Xyl, xylose; Gal, galactose; Ara, arabinose. Redrawn from Wilkie (1985).

would prevent their mobility. Flexible sidechains *hinder* crystallization as hemicellulose chains approach each other. This will permit the formation of an ordered but mobile cholesteric liquid crystalline phase, by allowing planar sheets of pipe-cleaner-shaped molecules to twist progressively through a small angle (Fig. 5.1B). This 'hairy helicoid' should be compared with the cellulose crystallite (Fig. 5.1A) in which the chains crystallize in an unimpeded manner in untwisted parallel array.

(3) As described in Section 4.3e, extracted hemicelluloses can be made to self-assemble *in vitro* into a cholesteric liquid crystalline phase (Roland et al., 1977; Reis, 1978).

(4) The deposition of plant cell wall matrix can be blocked by treatment with the ionophore drug *monensin* (Brummell & Hall, 1985). This drug has been used to prevent helicoidal wall formation in *Nitella*, despite the continued synthesis and deposition

of cellulose (Levy, 1986). This seems to implicate matrix hemicelluloses in helicoidal wall formation.

(5) For a given volume of cell wall, the matrix component is deposited first, and then the cellulose microfibrils are added to it (Northcote, 1977). This supports the idea that the matrix plays the key role in wall morphogenesis.

(6) When growing maize (*Zea mays*) was treated with the cellulose inhibitor DCB (dichlorobenzonitrile), xylose deposition in the cell walls increased at the expense of glucose (Satiat-Jeunemaître & Darzens, 1986). This graded response suggests that helicoid assembly is affected by the balance between cellulose and hemicellulose deposition in the periplasmic space. Subsequent work on mung beans supports this (Satiat-Jeunemaître, 1987). Cell wall deposited during treatment with the herbicide DCB lacks helicoidal structure. Although the cellulose is absent from such walls, hemicellulose is nevertheless still being deposited. This has been interpreted to mean that hemicelluloses are not by themselves able to self-assemble *in vivo* into a helicoid (Satiat-Jeunemaître, 1987). Cooperation of cellulose with hemicelluloses may be required for helicoidal wall formation.

(7) There is some correlation between the taxonomic distribution of hemicelluloses and helicoidal cell walls. Thus, xyloglucan is found in monocotyledons (Kato, Iki, & Matsuda, et al., 1981), in dicotyledons (Bauer et al., 1973), and in gymnosperms (Thomas, Darvill, & Albersheim, et al., 1983). Helicoidal walls are known from all these groups. By contrast, the algae *Valonia*, *Cladophora*, and *Chaetomorpha* have cell wall microfibrils of pure cellulose (i.e. without a sheath of hemicellulose). Helicoidal wall structure is absent from these plants.

(8) There is also a correlation between the localization of hemicelluloses and helicoidal layers in wood. A tracheid cell wall consists of three helical S-layers and two thin intervening helicoidal layers (Section 2.5b). If hemicelluloses are the organizers of helicoidal layers, they would be expected to be concentrated within them. The literature already contained some evidence for the concentration of glucuronoxylans between S-layers (Meier, 1961; Parameswaran & Liese, 1982). Elegant proof was obtained in Professor J.-C. Roland's laboratory in Paris (Vian et al., 1986). They labelled the enzyme xylanase with the heavy metal gold and exploited the specificity of the enzyme to locate and bind with xylose residues in the backbone chains of glucuronoxylan hemicelluloses (Fig. 5.3). Electron micrographs of sections through tracheid cells in the wood of linden trees (*Tilia platyphyllos*) showed the electron-dense gold-labelled xylanase to be concentrated in the helicoidal zone between S-layers (Fig. 5.4). The result was summarized in a diagram (Neville, 1988a), with the hemicelluloses represented as pipe-cleaner shapes (Fig. 5.5). Supporting evidence may be deduced from work by Takabe et al. (1984). They used

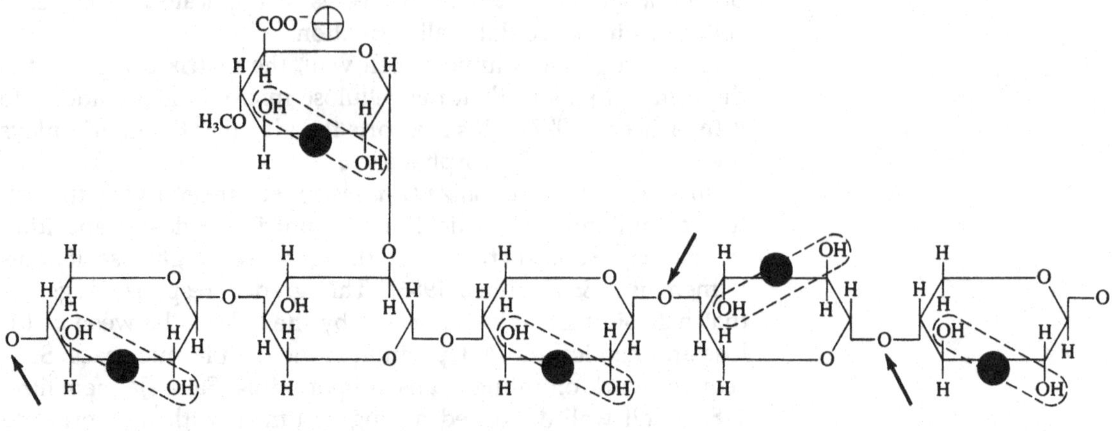

Figure 5.3. A small part of a hemicellulose molecule (glucuronoxylan) found in helicoidal plant cell walls. This is thought to be a prime organizer of helicoidal architecture. There is one residue per sidechain, every fifteen backbone xyloses. Sidechains are thought to be involved in polysaccharide cholesteric liquid crystal formation. The glucuronoxylan chains may be labelled with heavy metals, enabling their location in cell wall architecture to be found. Solid black circles indicate sites of action of periodic acid/thiocarbohydrazide/silver proteinate test (PAT Ag test); the circle with the plus sign indicates cationized iron binding; the arrows localize sites of xylanase-gold complex. Redrawn from Vian et al. (1986).

autoradiography to localize radioactively labelled glucose within xylan-rich polymers in the S_1–S_2 border of tracheids in wood from the conifer *Cryptomeria japonica*.

(9) Hemicelluloses are further implicated in helicoid formation by results obtained on the cholesteric liquid crystalline mucilage from quince (Section 4.3f). Hydrolysis and gas-liquid chromatography yielded uronic acids, xylose, rhamnose, fucose, arabinose, and galactose (Willison & Abeysekera, 1989). These workers suggested that the major chemical of the mucilage, which contains these components, is responsible for determining the cholesteric form of the periplasmic liquid crystal. This major branched polysaccharide has now been identified as (4-O-methyl-D-glucurono)-D-xylan (Lindberg et al., 1990). This seems like a potentially cholesteric substance, with stiff backbone, asymmetric carbons, and sidechains.

Evidence is therefore accumulating in support of the suggestion (Neville, 1985a; Vian et al., 1986; Satiat-Jeunemaître, 1987) that the hemicelluloses are the prime organizers of helicoids in plant cell walls. Helicoids are, however, visualized, in both light and electron microscopes, by the orientation of cellulose microfibrils. The final piece in the jig-saw is therefore to explain how hemicelluloses manipulate cellulose microfibrils into a helicoid.

Evidence from infra-red spectrophotometry of wood shows that hemicelluloses and cellulose microfibrils are oriented in parallel (Liang et al., 1960; Marchessault & Liang, 1962). The birefringences of hemicellulose and cellulose in coniferous wood are equal; this also indicates that they lie in parallel (Page et al.,

5.1 Self-assembly: hypotheses / 153

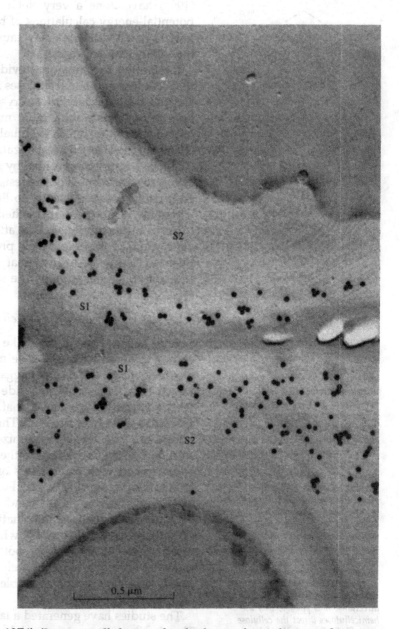

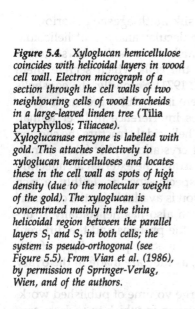

Figure 5.4. Xyloglucan hemicellulose coincides with helicoidal layers in wood cell wall. Electron micrograph of a section through the cell walls of two neighbouring cells of wood tracheids in a large-leaved linden tree (Tilia platyphyllos; Tiliaceae). Xyloglucanase enzyme is labelled with gold. This attaches selectively to xyloglucan hemicelluloses and locates these in the cell wall as spots of high density (due to the molecular weight of the gold). The xyloglucan is concentrated mainly in the thin helicoidal region between the parallel layers S_1 and S_2 in both cells; the system is pseudo-orthogonal (see Figure 5.5). From Vian et al. (1986), by permission of Springer-Verlag, Wien, and of the authors.

1976). Because cellulose and xyloglucans have the same $\beta(1 \rightarrow 4)$ backbone repeat, they are able to self-assemble (co-crystallize) by hydrogen-bonding together (Bauer et al., 1973; Albersheim, 1975). Their large sidechains restrict the hemicellulose sheath around each cellulose crystallite to a single molecule thick (Valent & Albersheim, 1974). Evidence in strong support of the idea that hemicellulose self-assembles to form a sheath around each cellulose crystallite has recently been obtained for quince mucilage (Reis et al., 1991). This is summarized in Section 4.3f. Levy et al.

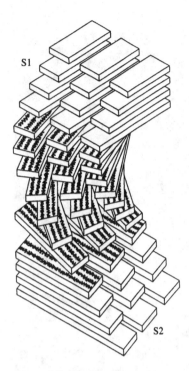

Figure 5.5. A pipe-cleaner molecular model relating chemistry to ultrastructure in wood cell walls. Each lath-shaped building unit represents a microfibril in the form of parallel cellulose molecules. In either the S_1 or the S_2 regions, the cellulose microfibrils lie in parallel, but are set at a large mutual angle. The S-layers spiral around a wood tracheid cell wall forming a tubular plywood. Hemicellulose molecules are represented as pipe-cleaner shapes to indicate their bulky and flexible sidechains. These hemicelluloses are part of the continuous matrix, and they also coat the surfaces of the cellulose microfibrils. These latter hemicelluloses coincide with the helicoidal regions which rotate between the S_1 and S_2 directions. It is possible that the hemicelluloses direct the cellulose microfibrils into a helicoidal array. From Neville (1988a), by permission of Academic Press Limited, London.

(1991) have done a very detailed study based on theoretical potential-energy calculations of binding xyloglucan to cellulose; a fucosylated sidechain may cause a wave of local flattening of the xyloglucan backbone.

Integrating the foregoing evidence, it is therefore suggested that cellulose and hemicelluloses interact to form a hemicellulose sheath around each cellulose crystallite. The sheathed crystallites then self-assemble with more matrix hemicelluloses to form a helicoidal texture, which eventually stabilizes. It is noted by Page (1976) that perfect co-crystallization between hemicellulose and cellulose would be prevented by steric hindrance from the hemicellulose sidechains. He suggests that in woody cells any loose ends could bond covalently to lignin.

Emons (1986) discounts the helicoidal self-assembly model because cellulose microfibrils are attached to the plasma membrane via terminal complexes (their probable site of polymerization). Even so, it can be argued that their free ends would still be available for orientation by the surrounding matrix.

5.1b Silk moth eggshells: proteins

Several factors combine to make silk moth eggshell chorion the chosen material for a full-scale molecular analysis of helicoidal morphogenesis. Firstly, the eggshells have helicoidal structure which is self-assembled outside the cells which secrete them (Telfer & Smith, 1970; Smith et al., 1971). Secondly, the eggshells are made entirely of proteins. There is no chitin present to complicate the system; this minimizes interference in laser Raman studies. Thirdly, in addition to the helicoidal structure the chorion is permeated with many kinds of crystallographic faults (Smith et al., 1971; Mazur et al., 1982). Fourthly, the external architecture of the eggs is complicated and species-specific. Fifthly, a rich background of genetical information is available for the commercial silk moth (*Bombyx mori*), which also has helicoidally structured eggshells. And sixthly, both the proteins and the different stages of secretion follow a well-defined sequence. Eggs of different stages are readily available, as they are made in the ovary on a production-line basis.

The studies have generated a large volume of published work. The chorion represents a model system in which to address several of the major problems in developmental biology: control of gene expression, evolution of multigene families, morphogenesis of supermolecular structures, structural protein folding and organization.

As many as 186 different proteins have been separated from the eggshell of *Antheraea polyphemus* by two-dimensional gel electrophoresis. These belong to six multigene families, divisible into two branches. Although the large number of proteins adds

5.1 Self-assembly: hypotheses / 155

complexity, they are all related, are strongly linked, and all map on chromosome two. Like other extracellular proteins, they initially possess a signal peptide sequence of about twenty amino acids, prior to export from the cells. They have fairly low molecular weights: class A (9–12,000), class B (12–14,000), class C (16–20,000). The C-proteins are responsible for the formation of the initial helicoidal framework. They are secreted early in eggshell formation, and ten to twelve genes are involved. In a mutant of *Bombyx mori* called G_r^{col} both C-protein secretion and helicoidal structure are absent (Nadel et al., 1980).

The proteins differ only by subtle changes in amino acid sequences (e.g. replacements, insertions, and deletions), having all derived from an ancestral gene (Regier & Kafatos, 1985). More than fifty of their primary sequences have been established by direct sequencing, or by sequencing the corresponding cloned DNA. It is, however, their secondary structure which is relevant to an understanding of their self-assembly capabilities.

The available evidence points towards a large amount of β-pleated sheet conformation, many β-turns, and a low α-helix content. Initial predictions were based upon analysis of primary sequences (Hamodrakas, Jones, & Kafatos, 1982b). X-ray diffraction indicated the presence of many β-pleated sheets oriented parallel to the chorion surface (Hamodrakas et al., 1983). Laser Raman spectroscopy confirmed a 60–70% content of anti-parallel β-pleated sheet (Hamodrakas et al., 1982a; Hamodrakas, Kamitsos, & Papanikolaou, 1984).

As mentioned in Section 3.1c, eggshell chorion proteins are tripartite; they have a central core region, with a flexible arm to either side (Fig. 3.5A). The sequence in the central cores shows little variation – it is highly conserved. It is likely that the cores bond together in groups to form the fibrous part of a helicoid, while the arms associate to form the matrix (Fig. 3.5B). The core accounts for about 20% of the total chorionic mass, which would be about right for a mechanically effective fibrous composite material. The arms are thought to be responsible for spacing out the fibrils. It may be significant that the arms of the C-proteins, which make the initial helicoid, are long and unique; perhaps they permit a cholesteric liquid crystalline phase to form (compare with the sidechains of hemicelluloses and cellulose derivatives described in Section 5.1a).

Computer analysis reveals a six-fold hexapeptide repeat in the sequence of the central cores. This usually takes the form of glycine-X-large hydrophobic-Y-large hydrophobic-Z. The residue (X) is a β-turn-former. A diagram of a typical central core, based on work by Hamodrakas et al. (1988), is shown in Figure 5.6. The positions of the hydrogen bonds which stabilize this β-pleated sheet are shown. Like the β-pleated structure in silk fibroin, these chorion proteins often have an alternation of small

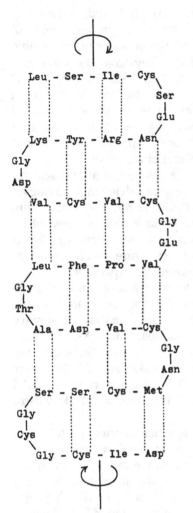

Figure 5.6. Chemistry and configuration of a typical moth eggshell protein. The amino acid sequence is typical for the central domain. This anti-parallel β-sheet model is based on the results of Hamodrakas et al. (1988). The turns involve a glycine. The hydrogen bonds have been drawn in (dotted). This β-sheet conformation is probably twisted, as in Figure 5.7. This is important, as Hamodrakas and associates propose such a propellor-shaped unit as the basis for helicoidal supramolecular structure in moth eggshell chorion.

sidechains (glycine, alanine, threonine) and larger sidechains (valine, leucine, isoleucine). The sidechains then pack more closely with those of the next sheet; large and small amino acids are grouped on opposite sides of the β-sheet.

Hamodrakas (1984) made the important proposal that the β-pleated sheet structure in chorion protein central cores was twisted and that the twisted β-pleated sheet was responsible for stacking with its neighbours to form a helicoidally twisted structure. It is known that most β-pleated sheets are twisted, with a left-handed rotation; for references, see Hamodrakas et al. (1988). The diagram in Figure 5.6 is drawn with arrows of torsion appropriate to generate a left-handed twist. Twisted β-pleated sheets have helical conformation.

As proteins do not branch, they do not have bulky flexible sidechains (unlike hemicelluloses). Perhaps the long flexible glycine-rich arms of the chorion proteins take on the task of the hemicellulose sidechains in plant cell wall helicoids, by hindering crystallization of the central cores? They could then twist through a small angle relative to their neighbours (see Section 5.1a).

Three-dimensional models of the twisted β-pleated sheets from central cores have been computed on the basis of stereochemical analysis (Hamodrakas et al., 1988). Stereo computer drawings of a twisted β-pleated sheet are reproduced by kind permission of Professor Stavros Hamodrakas in Figure 5.7. Two methods to appreciate the three-dimensional nature of this proposed structure are given in the legend. The structure, which resembles an aircraft propellor in shape, has a diameter of approximately 3.0 nm. This is in agreement with the 3.0-nm-diameter filaments seen in replicas of freeze-fractured surfaces (Hamodrakas et al., 1986).

The packing of such twisted β-pleated sheets to form higher-order structures is likely to play an important part in future studies of this and other helicoidal systems. It is proposed (Hamodrakas et al., 1988) that the packing of central cores into helicoids may follow the rules for packing twisted β-pleated sheets established by Chothia, Levitt, and Richardson (1977).

5.1c Fish eggshells: proteins

As with moth eggshells, the discovery of helicoidal structure in fish eggshells (Grierson & Neville, 1981) has led to studies on the secondary conformation of its component proteins (Hamodrakas et al., 1987). It is of interest to enquire whether there are any basic rules to helicoid assembly which are common to these unrelated systems. The work summarized in Section 5.1b indicates the probable underlying importance of twisted β-pleated sheet secondary conformation in helicoid formation in moth

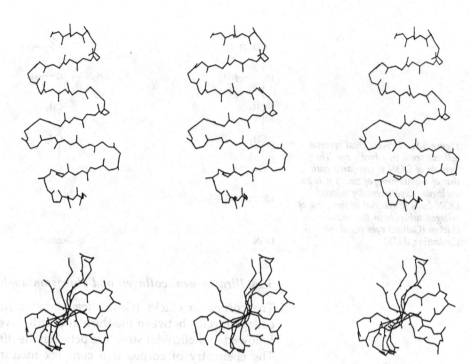

Figure 5.7. The probable molecular basis of helicoid formation in moth eggshell proteins. Complementary DNA sequences of moth chorion proteins were converted into protein amino acid sequences. The conservative central domain of a typical protein is shown. Interactive graphics were used to give a model of the characteristic β-pleated sheet; only the main polypeptide chain and its projecting carbonyl oxygens are shown. The plane of the β-sheet is twisted; this can be seen either by using a stereoscopic viewer or, if one is so gifted, by crossing one's eyes. The authors recommend using the left and middle images for normal vision, and the middle and right-handed images for cross-over vision. It works best for me if I put the plane of the illustration about 15 cm from the tip of my nose, cross my eyes, and stare at the space between the two selected images. I can then see a blurred image to either side, and a sharper central image. After a few seconds delay the central image sharpens in 3-D, enabling the sense of twist of the computer model to be determined (clockwise moving away from the observer). It is worth persisting with this method of viewing even if it does not work the first time; when achieved, it is very satisfying. It does not, however, work for everybody. The top three images view the twisted β-pleated sheet sideways-on, whereas it is seen end-on in the lower three images. The conformation is about 3 nm in diameter. It is likely that the twisted β-pleated sheets of neighbouring proteins stack to form a helicoid. A similar 3-D effect is seen in the helicoid in Figure 1.9. Reprinted from Hamodrakas, et al. (1988), by permission of Oxford University Press and the authors.

eggshells. Hamodrakas et al. (1987) have used infra-red and laser Raman spectroscopy to investigate eggshells from rainbow trout (*Salmo gairdneri*); these are known, from the patterns made by both microfibrils and radial canals, to be helicoidal (Grierson & Neville, 1981). The results indicate an abundance of anti-parallel β-pleated sheet conformation. Taken together with the absence of α-helical conformation, this secondary structure should determine how the protein molecules interact to form helicoids. Corroboration by amino acid sequencing is required, but the similarity to the secondary structure of helicoidal moth eggshell proteins is striking.

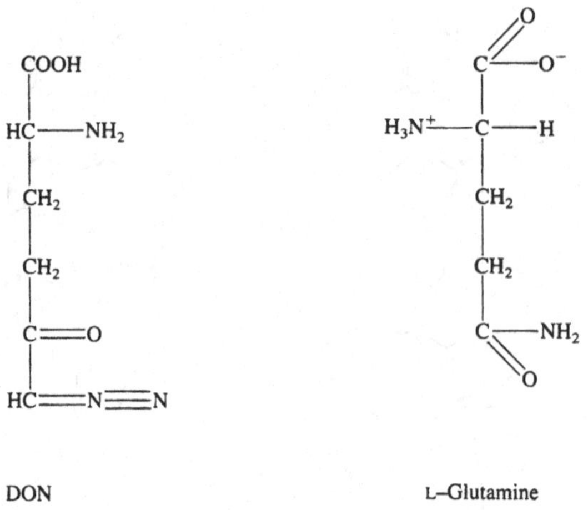

Figure 5.8. Experimental reversal of helicoid sense in a bird's eye. The formula of DON is compared with that of L-glutamine, of which it is an analogue. Experimentally injected DON leads to reversal of the sense of collagen helicoids in the cornea of chicken (Gallus) eyes (Coulombre & Coulombre, 1975).

5.1d Bird cornea: collagen and glycosaminoglycans

The cornea in chicks (*Gallus domesticus*) forms an orthogonal collagen trellis between the third and fifth days of development, followed by helicoidal structure between the fifth and tenth days. The chemistry of cornea is a complex mixture of protein and polysaccharide. Which molecules are important in forming this architecture?

Collagen can self-assemble into a helicoid in cell-free preparations (Giraud-Guille, 1989); it is also competent to form helicoids in the absence of polysaccharides. However, polysaccharides are known to have an important influence on corneal morphogenesis. Coulombre and Coulombre (1975) treated five-day-old chicks with the antibiotic DON (6-diazo-5-oxo-L-norleucine). This is an analogue of L-glutamine (Fig. 5.8) and is able to inhibit the synthesis of glycosaminoglycans. This drug suppressed the formation of normal helicoidal architecture and even appeared to reverse the usual sense of rotation (i.e. from right- to left-handed in both eyes). These results have been confirmed in our laboratory by J. W. Dodson and D. C. Gubb and are reproduced by their permission in Figure 5.9A,B. So polysaccharides play an important part in the final architecture, despite the intrinsic capabilities of collagen.

It has been shown that *chondroitin sulphate* is not the molecule which organizes orthogonal collagen layers (Bansal, Ross, & Bard, 1989), despite being the most abundant proteoglycan in vertebrate extracellular matrix. Bansal and co-workers grew chick corneal stroma in organ culture from which chondroitin was continually removed by a specific enzyme; their results support self-assembly of orthogonal collagen. Similarly, Hay (1980) reasons that *keratan sulphate* is not responsible for the morphogenetic control of orthogonal collagen in chick corneal primary (non-

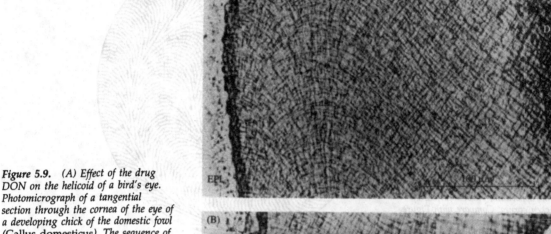

Figure 5.9. (A) Effect of the drug DON on the helicoid of a bird's eye. Photomicrograph of a tangential section through the cornea of the eye of a developing chick of the domestic fowl (Gallus domesticus). The sequence of layer deposition in the stroma (S) runs from the endothelial (END) side to the epithelial (EPI) side. Collagen fibres are stained with Gomori's silver technique. Descemet's membrane (DM) is also shown. In this control experiment, the embyo was injected with 5 μl of water between 4.5 and 5 days old, and fixed 5 days later. The later layers show helicoidal layer rotation (H). (B) As for part A, but injected instead with 29 μmol of DON in 5 μl of water. There is an apparent halt, or even a reversal (R), in the sense of layer rotation. By kind permission of Dr. J. D. Dodson and Dr. D. C. Gubb, whose results confirmed those of Coulombre and Coulombre (1975).

cellular) stroma, because it does not contain this polysaccharide.

When cut tangentially, chick cornea produces a complex double spiral pattern, illustrated diagrammatically in Figure 5.10. This is created by an orthogonal pair of layers which acts as the building unit, each pair being set at a slight angle to its neighbour. Despite the large volume of biochemical information on extracellular matrix, very few workers are conscious of its complex fibrillar architecture.

5.1e Sense of helicoidal twist

The sense of rotation of helicoids in biology is constant, with very few exceptions. This is to be expected if they arise via a cholesteric liquid crystalline phase, whose sense of rotation is in

Figure 5.10. Diagram of the double spiral pattern arising in an equatorial section through a helicoidal spherulite with an angle of 89° between successive layers. (This is a spherical twisted plywood construction; see Fig. 4.6.) This type of pattern, with overlapping arcs, is seen in sections of the spherically curved surface of chicken cornea. Compare with Figure 5.9A. By kind permission, from the Ph.D. thesis of Dr. D. C. Gubb (1976).

turn determined by L or D optical rotation – itself the result of stereochemistry. Systems for which we have information are given later. The electron microscope is usually needed, and the procedure for determining sense of rotation is given by Neville and Luke (1971a). All of the examples have left-handed twist; that is, the layers rotate anti-clockwise in steps going farther away from the observer. Several arthropod cuticles have been investigated, such as the shore crab *Carcinus maenas* (Bouligand, 1965), the Pacific prawn *Panulirus argus* (Neville & Luke, 1971a), the freshwater crayfish *Astacus fluviatilis* (Neville, 1975a), the giant water bug *Belostoma malkini* (Neville, 1975a), both larval and adult mealworm beetles *Tenebrio molitor* (Neville, 1975a), and the golden scarab beetle *Plusiotis resplendens* (Caveney, 1971). To this list should be added several thousand species of adult scarab beetles whose rotation was determined optically (Section 3.3b) by circular analyzer (Neville & Caveney, 1969). The cellulose test of the sea squirt *Halocynthia papillosa* also has left-handed helicoidal structure (Gubb, 1975), as do the eggshells of some fish: cod (*Gadus morrhua*), plaice (*Pleuronectes platessa*), and rainbow trout (*Salmo gairdneri*) (Grierson & Neville, 1981). These chordate examples have helicoids of such large pitch that their sense of rotation may be determined from light microscopy of arced patterns in suitably oblique sections. It should be noted that the sense of rotation of the pseudo-orthogonal layer system in the scales of fish is dependent upon species. The sense of twist in the lemon sole *Pleuronectes microcephalus* is right-handed (Fig. 2.13), whereas that of the roach *Rutilus rutilus* (Fig. 2.12) is left-handed (Darke, 1986).

5.1 Self-assembly: hypotheses / 161

Figure 5.11. *The unexplained reversal of helicoid every 180° in an external tube secreted by a deep-sea worm (Alvinella pompejana). Redrawn from Gaill and Bouligand (1987), who tentatively suggested that movements of the worm in its tube could cause this effect. Such reversals are not supposed to be seen in structures of liquid crystalline origin; in fact, it is believed that such regular reversals are forbidden.*

We also have information for helicoidal plant cell walls of the stonewort *Nitella opaca* (Levy, 1987), the alga *Boergesenia forbesii* (Mizuta & Wada, 1981), and the water plant *Limnobium stoloniferum* (Pluymaekers, 1982), all of which twist left-handed.

An exceptional example is the cylindrical wall of the tube secreted by the deep-sea worm *Alvinella pompejana*, whose helicoidal layers of fibres appear to reverse their sense of twist (Fig. 5.11) every 180° (Gaill & Bouligand, 1987). As this type of recurrent reversal was then unknown in liquid crystals, Gaill and Bouligand resorted to the proposal that motions of the worm inside its tube might cause the formation of this structure. However, this kind of structure has since been seen in a liquid crystal reassembled from cellulose microfibrils and glucuronoxylans, extracted from quince (*Cydonia*) mucilage (Reis et al., 1991). This offers a less embarrassing explanation than a wriggling-worm hypothesis. Another example of helicoid pattern reversal was published in my book (Neville, 1975a); this was of membranous cuticle in a locust skeletal joint.

5.1f Remote control of construction

The structures which form the subject of this book are assembled outside the cells which manufacture their components. Therefore, any mechanisms which enable the cells to exercise control must be remote. Possible types of mechanisms are pH, concentrations of other ions, polymer concentration, and water composition, as previously suggested (Neville, 1967c). A good example of the influences of all these factors on conformation is that of xanthan (Dentini, Crescenzi, & Blasi, 1984).

There is direct evidence for the effect of pH on helicoid self-assembly in two systems: mantis oothecal proteins (Gubb, 1976) and plant cell wall hemicelluloses (Reis, 1978). Mantis eggcase proteins may be freeze-dried and reassembled into a helicoid in sodium sulphate solution. A phase change occurs at pH 5.0; above 5 it is helicoidal, and below 5 isotropic.

That a plant cell can in fact control extracellular pH was shown for *Avena* by Rayle (1973). Auxin (indole acetic acid) acts on the cell to control pH concentration in its cell wall. This, in turn, regulates wall extensibility (Fig. 5.12A).

A similar mechanism is known for the epidermal cells of the blood-sucking bug *Rhodnius* (Reynolds, 1975). Here 5-HT (5-hydroxytryptamine) acts on the epidermal cells to make them change the pH out in the neighbouring cuticle. The change from pH 7.0 to pH 5.0 breaks the double bonds in Schiff bases which previously held the cuticle protein chains together. The cuticle then becomes sufficiently extensible (Fig. 5.12B) to cope with the large volume of blood ingested when the bug feeds. The remarkable similarity between these two examples of remote control

162 / *How is fibre orientation controlled?*

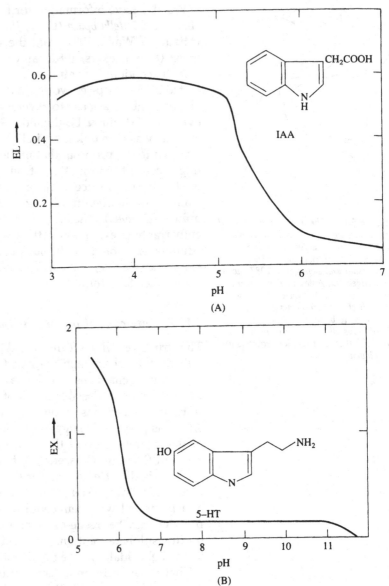

Figure 5.12. Similarity of control of extensibility of a plant cell wall and the cuticle of an insect. (A) Graph showing how extensibility in a plant cell wall (Hordeum) under constant strain is affected by the hormone auxin (IAA, indole acetic acid). This controls pH in the cell wall, which in turn controls cell elongation, measured in millimetres/hour. Redrawn from Rayle (1973). (B) The cuticle of the insect *Rhodnius* has to become very extensible to cope with the massive aqueous volume of its food – it feeds on vertebrate blood. The extensibility under constant strain is affected by 5-HT. This controls pH in the cuticle, which in turn controls its elongation (in millimetres/minute). Redrawn from Reynolds (1975). Note the remarkable similarities with part A.

mechanisms has been noted before (Neville, 1985b). This extends even to the chemical similarity of indole acetic acid and 5-HT. We do not yet know if cells control helicoid self-assembly by pH regulation, but it is feasible.

5.2 Directed assembly: hypotheses based on cellular mechanisms

This section summarizes various proposed cellular mechanisms which may be involved in control of fibre orientation. These are not necessarily to be viewed as alternatives to the molecular self-assembly mechanisms presented in Section 5.1.

5.2a Microtubules and microfibrils

The advent of glutaraldehyde as a fixative for electron microscopy made possible the discovery of microtubules, because it prevents them from depolymerizing into tubulin monomers. Microtubules are fixed by glutaraldehyde molecules because each has two aldehyde groups, as opposed to the single group of formaldehyde. Ledbetter and Porter (1963) discovered microtubules in plant cells and proposed that those of the cortical zone (lying immediately beneath the plasma membrane) were involved in controlling the orientation of extracellular cellulose microfibrils, because the two systems were often observed to lie in parallel. A summary of evidence for this view is given by Heath and Seagull (1982). In his 1982 book, and in a 1983 review, Lloyd has marshalled compelling evidence that microtubules and cellulose microfibrils often lie in parallel with each other to either side of the cell plasma membrane. The implication is that microtubule orientation controls microfibril orientation. A recent textbook (Alberts et al., 1989) cites further examples. One is in the developing spiral thickening of xylem cells; another is in the cotton hair cell wall, in which the orientation of microtubules (revealed by immunofluorescence) is shown to lie parallel to that of cellulose microfibrils (revealed by fluorescence using calcofluor white).

There are, however, exceptions: In the alga *Valonia* there is no correspondence between the directions of cortical microtubules and the most recently deposited cellulose microfibrils (Itoh & Brown, 1984). In another alga, *Chamaedoris orientalis*, the newest microfibrils lie at various angles to the filament axis (transverse, oblique, longitudinal), whereas the cortical microtubules lie always in the direction of the axis (Mizuta et al., 1989). One heading in the textbook by Alberts et al. (1989) – which contains an otherwise excellent account of extracellular materials – states that cellulose microfibrils are deposited parallel to cortical microtubules. This, while true for some plant cell walls, is misleading for very many others – namely, those with helicoidal walls. We have expressed the view (Neville & Levy, 1984) that directed assembly of helicoids of microfibrils by cortical microtubules is highly unlikely, because it would need a continual sequence of disassembly of tubulins, followed by their reassembly in a slightly new direction. This would demand hundreds of alternate depolymerizations and repolymerizations of microtubules.

In the example of the green alga (*Boergesenia forbesii*), helicoidal wall is deposited in the complete absence of cortical microtubules at any stage of development (Mizuta & Wada, 1982). In the walls of root hairs of the horsetail *Equisetum hyemale*, the cortical microtubules are oriented constantly along the stem axis throughout the deposition of several turns of helicoidally oriented cellulose microfibrils (Emons, 1982).

By means of immunofluorescence, Satiat-Jeunemaître (1989) has visualized the orientations of all the cortical microtubules in hypocotyl cells of germinating mung beans (*Vigna*). In the young cells, clear helicoidal microfibrillar architecture is seen, while the cortical microtubules lie mostly transverse to each cell long axis. In slightly older cells, helicoids are still seen, but now the microtubules are mostly oblique to each cell axis. In yet older cells the helicoidal patterns are undergoing dissipation, while the microtubules lie mostly along the long axis of each cell. There is no coincidence between microfibrils and microtubules.

Taking a wider view, microtubules are not present during deposition of the cyst wall of the protistan *Actinophrys sol* (Newman, 1990). The cyst wall contains a helicoidal layer. In animal cells in general, many of which secrete helicoidal layers of microfibrils, the cortical zone does not contain microtubules. Neither, of course, are microtubules present during the self-assembly of helicoids *in vitro*, as described in Chapter 4. It is concluded that microtubules are unnecessary for helicoid formation; but, as we have emphasized (Neville & Levy, 1984), this does not exclude their possible involvement in those plant cell walls (and other systems, such as the chitin in insect hairs) which are not helicoidal. In a recent review, Preston (1988) also concluded that the range of microfibrillar architectures found in plant cell walls cannot be accounted for by microtubule mechanisms. It was disappointing to see no mention of helicoids in that paper.

5.2b Terminal-complex rosettes: a microtubule channel model

A problem with any model which links microtubules with microfibril orientation is that one is inside and the other outside the cell plasma membrane. How are they connected? Freeze-fracture techniques developed over the past decade enable the membrane to be split into two parts, revealing *terminal complexes* in the outer (E) face of the membrane, and *rosettes* in the inner (P) face of the membrane. Mueller and Brown (1980) proposed that microfibril orientation in plant cell walls is controlled by these complexes, moving in the plane of the membrane and depositing microfibrils in their wake. The Singer and Nicolson (1972) fluid mosaic model for membranes recognizes the smectic liquid crystalline nature of the membrane lipids. This makes possible movement of organelles like the rosettes in the plane of the plasma membrane.

It is reasonably established that the rosettes seen in the plant plasma membrane are the sites of origin of cellulose microfibrils. As a microfibril lengthens, so its rosette is pushed in the opposite direction. Giddings and Staehelin (1988) proposed a model to connect the orientation of intracellular cortical microtubules with extracellular cellulose microfibrils. They envisage cross-bridges

Figure 5.13. Possible model linking microtubule orientation and cellulose microfibril orientation in plant cells. Cellulose microfibrils are synthesized on rosettes, which are thought to move in the plane of the fluid mosaic plasma membrane. The forces which may move them are the polymerization and secretion of cellulose. As a microfibril grows, the rosette is pushed in the opposite direction. The model attempts to link cortical microtubules (inside cells) to cellulose microfibrils (outside cells), via linking cross-bridges. The authors of this model propose that the cellulose synthetase rosettes are channeled to move only between the constraints of two microtubules. This results in cellulose microfibrils being laid down in parallel with the cortical microtubules. ROS, rosette; MEM, cell plasma membrane; MIC, microtubule; LINK, membrane–microtubule cross-bridge link. Redrawn from Giddings and Staehelin (1988).

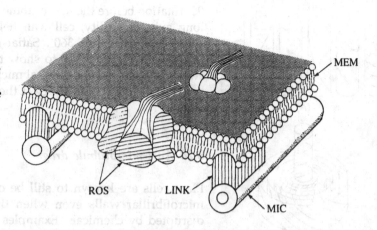

linking microtubules to the plasma membrane (Fig. 5.13). The rosettes are supposed to be motile, but channeled between microtubules, like a ship is channeled by the banks of a canal. The result is that the microfibrils are laid down parallel to the microtubules. In support of this model, Giddings and Staehelin (1988) found that in the single-celled alga *Closterium*, rows of rosettes occur between microtubules rather than on top of them. The motor force driving a rosette is thus likely to be the force of cellulose polymerization, rather than a force generated by interaction of rosette-bound proteins sliding along a microtubule. Another suggestion is that dynein powers the terminal rosettes (Heath & Seagull, 1982).

5.2c Shifting multistart helices of microtubules

This model was proposed by Roberts, Lloyd, and Roberts (1985) in response to my criticism (Neville, 1985a) that a microtubule model for generation of helicoidal walls was unlikely; it would require a depolymerization and a repolymerization of the microtubules for each angular shift of the microfibrils. In root hair tips and apical cells of a moss, *Physcomitrella patens*, both cortical microtubules and the innermost layer of wall microfibrils are wound helically around the cylindrical wall. The course of the microtubules was revealed by staining with fluorescent anti-tubulin (Roberts et al., 1985). When treated with ethylene, the microtubular helices changed from low through medium to long pitch, showing that the whole array of cortical microtubules consists of a single multistart helix which can adopt different degrees of winding, like a spring (Fig. 5.14A). The authors suggested that this change in microtubule winding angle could generate a helicoid of wall microfibrils, provided that the microtubules and most recent layer of microfibrils were aligned in parallel. This model was rejected (Neville, 1985a) because it is capable of only

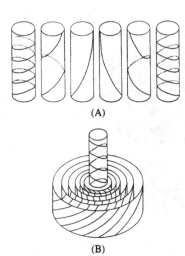

Figure 5.14. Can microtubules influence helicoids in plant cell walls? A model has been proposed which links systematic changes in direction of cellular cortical microtubules with the helicoidal rotation of extracellular microfibril orientations (Roberts et al., 1985). (A) Helical distribution of microtubules within an elongating cylindrical plant cell wall. For simplicity, only one microtubule is shown; immunofluorescence studies show that a cortical field of parallel microtubules follow such a helical path around the cell. At any one time, the innermost layer of microfibrils is supposed to follow the same helical course as the cortical microtubules. Ethylene induces the helical field of cortical microtubules to change position, like an alternately extending and rewinding spring. This is supposed to give a continued clockwise rotation of microtubules, as shown in the sequence from left to right in part A. I have criticized this model (Neville, 1985a) on the grounds that an extending spring cannot produce a rotation of more than 90°; yet most plant cell wall microfibrillar helicoids rotate through several turns of 360°. Redrawn from Roberts et al. (1985). (B) A helicoidal sequence of wall microfibrils arranged concentrically around a cylindrical plant cell which contains helically wound cortical microtubules. Satiat-Jeunemaître drew this figure to illustrate the hypothesis of Roberts et al. (1985), but, like me, she could not see how it could work. Redrawn from Satiat-Jeunemaître (1989).

90° rotation before the microtubules become aligned with the cell long axis. In reality, cell wall helicoids make many complete turns through a full 360°. Satiat-Jeunemaître (1989) drew the diagram in Figure 5.14B to show the proposed relationship between helically wound cortical microtubules and a helicoidal cell wall. This led her also to reject the explanation of Roberts et al. (1985).

5.2d Anti-microtubule drugs

Plant cells are known to still be capable of forming helicoidal microfibrillar walls even when their microtubules have been disrupted by chemicals. Examples are maize (*Zea mays*) treated with colchicine or with isopropyl-N-phenyl carbamate (Satiat-Jeunemaître, 1984), and the filamentous green alga *Chamaedoris orientalis* treated with the herbicide amiprophos-methyl (APM) by Mizuta et al. (1989). In the case of *Oocystis solitaria* treated with methylbenzimidazol-2-yl carbamate (MBC), helicoids are formed even where more or less orthogonal microfibril architecture is normally deposited (Robinson & Herzog, 1977).

In some cases, anti-microtubule drugs lead to the deposition of large arced patterning (long-pitch helicoid), where normally small-pitch helicoids would be deposited. Examples are mung bean (*Vigna radiata*) treated with colchicine (Vian et al., 1982), maize (*Zea mays*) treated with either colchicine or carbamate (Satiat-Jeunemaître, 1984), and stonewort (*Nitella opaca*) treated with colchicine (Levy, 1986). Large-pitch helicoids are illustrated for colchicine- and carbamate-treated maize cell walls, together with a control, by permission of Dr. Satiat-Jeunemaître (Fig. 5.15).

Several interpretations of these results are possible. Firstly, helicoidal wall is still produced after the disruption of microtubules. I would interpret this as evidence that microtubules are not necessary for helicoid formation. Secondly, there is a simultaneous effect upon microtubules and microfibril orientation, and some authors would interpret this as evidence for direct control of microfibril architecture by microtubules. An alternative explanation is possible, however. Northcote (1977) showed that hemicelluloses are synthesized in the Golgi apparatus, transported by membrane flow of vesicles to the plasma membrane, and exported through it by exocytosis. Because this transport is driven by microtubules, their disruption by drugs could reduce incorporation of hemicelluloses into the cell wall. Such a reduction in hemicellulose concentration could decrease the rate of rotation of wall helicoids. Pitch is known to vary inversely with concentration in cholesteric liquid crystals (Robinson, 1958). So the pitch of the wall helicoid could be enlarged, explaining why the arcs seen in Figure 5.15B,C are much larger than normal (Fig. 5.15A). Dia-

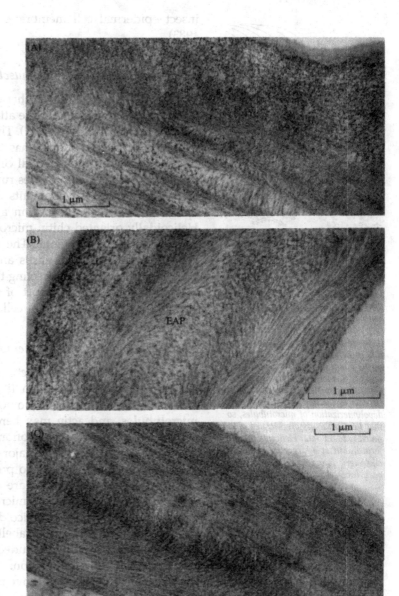

Figure 5.15. Effects of chemicals on helicoidal pitch in plant cell walls. (A) Electron micrograph of a section through the cell wall of the external epidermis of the growing coleoptile of maize (Zea mays; Graminaceae). The wall is helicoidal, but there is a loss of order (dissipation), accompanied by thinning of the outermost (= earliest) layers of helicoid. The micrograph shows control material. Reproduced from Satiat-Jeunemaître (1981), by permission from Masson, Paris, and the author. (B) As for part A, but treated with carbamate (isopropyl-N-phenyl carbamate) during wall deposition. This results in an enlarged arc pattern (EAP), indicating larger pitch and smaller angle of helicoid rotation; compare with the control in part A. Carbamate causes depolymerization of microtubules. This is thought to interfere with the transport of hemicellulose matrix through the epidermal cells to the cell walls. If hemicellulose is a key agent in helicoid formation, any reduction in its concentration could give rise to helicoids of greater pitch. See also the diagrams in Figure 5.16. Reproduced by permission from Satiat-Jeunemaître (1984). (C) As for part B, but treated with colchicine. This also disrupts the microtubular transport of hemicelluloses, producing a larger than normal arced pattern (see the explanation offered in part B). Parts B and C reproduced from Satiat-Jeunemaître (1984), by permission from Editions Scientifiques Elsevier, Paris, and from the author.

grams are given to summarize the control (Fig. 5.16A) and colchicine-treated (Fig. 5.16B) situations. So the effect of anti-microtubule drugs on cell wall microfibril orientation could be indirect, acting by disruption of the matrix transport system. It is known that helicoidal architecture may be restricted to certain regions of a plant cell wall (e.g. in charophyte oosporangia) (Leitch, 1986); this could result from regulation of matrix transport by microtubules to specific parts of a cell perimeter. There is a similar suggestion that microtubules may be involved in the movement of *chitosomes* (which make chitin) to exact sites in the

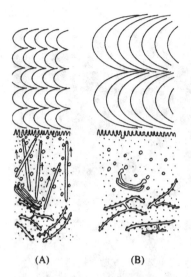

Figure 5.16. Diagrams to illustrate the effect of colchicine treatment on helicoid deposition in plant cell walls. (A) Without colchicine, hemicelluloses (believed by several of us to play a key role in helicoidal assembly) are formed in the Golgi and are transported in vesicles (arrows) by microtubules, to be exported from the cell by exocytosis. (B) Colchicine has caused the depolymerization of microtubules, so that hemicelluloses are secreted at a much-reduced rate. Helicoids are still formed, but at a larger pitch. Pitch in cholesteric liquid crystals is known to vary inversely with concentration.

insect epidermal cell membrane (Oberlander, Lynn, & Leach, 1983).

5.2e Microtubules in insect muscle attachments

In the case of insects (and other arthropods) the orientation of microtubules at sites of muscle attachment is *perpendicular* to the cuticular microfibrils (Fig. 5.17). The muscles pull on the cuticular exoskeleton, but their force has to be transmitted through the intervening layer of epidermal cells. These are strengthened by great numbers of microtubules running in the same direction as the actin and myosin filaments of the muscle (Fig. 5.18). The epidermal cells in this region are in fact called *tendon cells*. Helicoidally oriented chitin microfibrils are found in the cuticle at muscle insertion regions. The microtubules run perpendicularly to the chitin microfibrils and would not therefore be expected to play a part in directing their orientation, although they may be involved in transport of matrix proteins to the outside membranes of the epidermal cells.

5.2f Actin and microfibril orientation

It has been proposed that actin filaments in the animal cytoskeleton may orient collagen fibrils via fibronectin (Hynes & Destree, 1978). Animal cells do not possess a cortical network of microtubules, and actin may here be taking over their role in control of extracellular microfibril orientation. Actin was first discovered as one of the major proteins of muscle, but was subsequently found to be also present in plant cells. Both actin filaments and microtubules are oriented axially in root hairs (Derksen, 1986) but cellulose microfibrils (e.g. in *Ceratopteris* and *Limnobium*) are arranged helicoidally. Emons (1986) visualized fibrous actin by rhodamine-labelled phalloidin in cell walls of *Equisetum hyemale*, but concluded that it was not involved in cellulose microfibril orientation.

An example where cell movement causes orientation, in this case of collagen fibrils, is shown by fibroblast cells grown on a gel with random fibre orientation. The cells exert force on the substrate (fibroblast traction), creating ridges which orient the collagen (Harris, Stopak, & Wild, 1981).

5.2g Ordered-granule hypothesis

In plant cells the polymerization step from glucose to cellulose takes place in the terminal complexes sited in the plasma membrane. These complexes were previously known as cellulose synthetase granules. In cases where the granules were close-packed (Fig. 5.19), they were thought to be responsible for pro-

Figure 5.17. Diagram to show that microtubules are not involved in helicoid formation in insect cuticle. In insects the epidermal cells intervene between muscle and cuticle. The strain on the epidermis is borne by large numbers of microtubules (see Fig. 5.18), oriented in series with the muscle filaments and perpendicularly to the planes of chitin microfibrils in the cuticle. Helicoidal architecture occurs in the cuticle overlying all insect muscle insertions (apart from where they insert on tendons), despite the perpendicular disposition of the epidermal microtubules. We may conclude that microtubules are not necessary for helicoid formation in insect cuticle, though it is possible that they are involved in transport of cuticular components. CUT, cuticle; EPID, epidermis packed with microtubules; MUS, muscle.

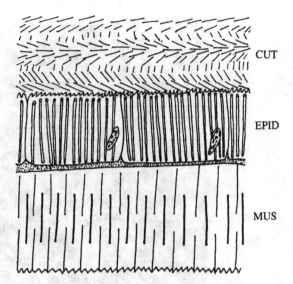

ducing cellulose chains oriented in specific directions (Preston, 1964; Preston & Goodman, 1967). We have commented (Neville et al., 1976) that whereas this model may explain some algal cell walls in which the angle between cellulose layers is 60° or 120° (hexagonal close-packed granules) or 90° (tetragonal close-packed granules), it could not account for helicoidal systems with much smaller changes in angle. In extreme cases the helicoidal angle can be as small as 6°, as, for example, in the alga *Boergesenia*, calculated from Mizuta and Wada (1982).

5.2h Circadian clocks and chitin orientation

The layers of chitin microfibrils in insect endocuticles are arranged either in a helicoidal manner or in a unidirectional bundle. This is the basis of the two-system model of cuticle (Section 1.3o). These two types of architecture are thought to arise by molecular self-assembly. However, the timing of their deposition is thought to be under the control of a cellular mechanism; circadian clocks cause the formation of daily cuticular growth layers.

A scanning electron micrograph of a piece of grasshopper cuticle (*Romalea microptera*) shows the location of the daily growth layers (Fig. 5.20). In early experiments on these (Neville, 1965a, 1967b), they were known as lamellate night layers and non-lamellate day layers. Subsequent work (Neville & Luke, 1969a) showed that these were equivalent to helicoidal chitin layers deposited at night and unidirectional chitin layers deposited during the day. The main experiments are described in the light of this further information.

If adult locusts (*Schistocerca* or *Locusta*) are reared in constant

170 / *How is fibre orientation controlled?*

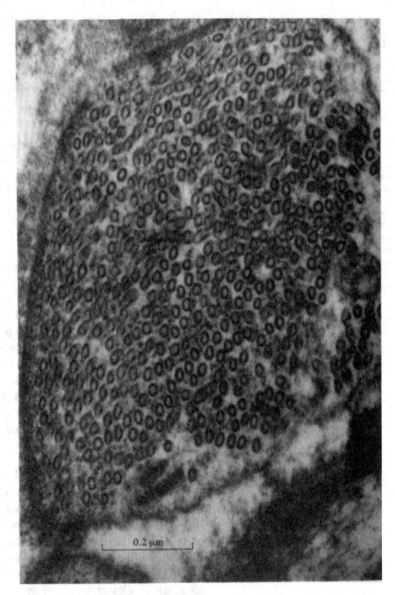

Figure 5.18. Electron micrograph of a section through the region of attachment of a muscle onto the hind femur of a one-day-old adult giant water bug (Hydrocyrius colombiae; Insecta, Hemiptera). The epidermal cells intervene between muscle and cuticle and hence need to be strengthened to take the force of the muscle when it contracts. This is done by filling the epidermis in such regions with microtubules – oriented in continuation of the direction of muscle fibres and perpendicularly to the layers of cuticle. The cuticle at muscle insertions contains helicoids, showing that their formation does not require microtubules oriented parallel to either the cell plasma membrane or the chitin microfibrils outside it (see Fig. 5.17). Micrograph by B. M. Luke and A. C. Neville.

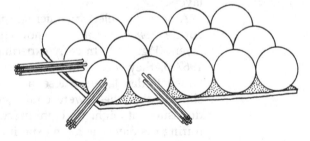

Figure 5.19. The oriented-granule hypothesis for control of plant cellulose microfibril orientation proposes that the orientations are determined by the packing of cellulose synthetase granules (fifteen are shown) on the outer face of the plant cell plasma membrane. The three possible directions of bundles of cellulose microfibrils are shown. This does not provide sufficient directions for the formation of a helicoid; in fairness, the model predates the recognition of plant wall helicoids. Redrawn from Preston (1964).

Figure 5.20. Insect exoskeletons document the history of their growth. This scanning electron micrograph shows a small block of cuticle cut from the hind tibia of a large adult grasshopper (Romalea microptera; Insecta, Orthoptera). The thin waterproof waxy epicuticle (EPI) records the polygonal outlines of the cells which made it. Below this lies the thicker laminated part, which is skeletal in function. The part made before the final moult is the exocuticle (EXO); the major part, deposited after the moult, is the endocuticle (ENDO). Its layers, which are deposited in a daily manner (D), record their own sequence of formation; they can be counted (Neville, 1963a) to give the exact age in days of the individual insect. From Hughes (1987), by permission of Churchill Livingstone Journals, Edinburgh, U.K., and of the author.

dark at 35 °C, their circadian clocks free-run with a twenty-three-hour rhythm, so that they are exactly out of phase with the natural environment after twelve days. They are then depositing helicoidal layers in what would have been the day, instead of at night. Likewise, they deposit unidirectional layers in what should have been night-time.

Other environmental conditions can override the circadian clock. Constant light at constant temperature (35 °C) for sixty hours leads to deposition in some parts of the exoskeleton of an abnormally wide band of unidirectional cuticle (Fig. 5.21), and if adult locusts are reared exclusively in these conditions, the whole endocuticle has unidirectionally oriented chitin microfibrils. By contrast, constant dark at a temperature of 25 °C leads in certain

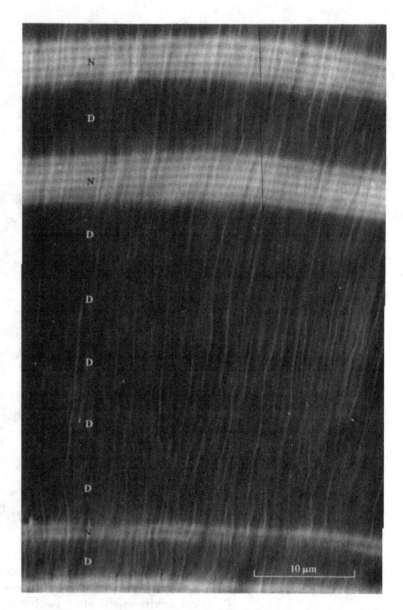

Figure 5.21. Experimental alteration of chitin architecture in insect cuticle. Light micrograph of a section through a hind tibia of an adult desert locust (Schistocerca gregaria; Insecta, Orthoptera), viewed between crossed polarizing filters. The cuticle is normally laid down in daily layers. The night layers (N) have helicoidal chitin architecture which creates a birefringent (white) layer for every half-turn through 180° (see Fig. 1.20). The day layers (D) have their chitin microfibrils oriented along the leg, so that they appear non-birefringent (black) in cross-section. The timing of the daily layers giving this change in chitin orientation is driven by a circadian clock. This can be overridden by environmental conditions, such as constant daytime (continuous light plus 35°C). The photograph shows a sequence of normal night (five half-turns), normal day (parallel), normal night (six half-turns), and then sixty hours of continuous day, giving a very wide (parallel) day layer (DDDDD). Normal conditions were then resumed to give a night-and-day alternation. In addition, pore canals can be seen running across the cuticle. These each contain a filament (transporting wax to repair the waterproof cuticle surface, like the wick in a candle). The filaments are only weakly birefringent. From Neville (1965a), reprinted by permission of the Company of Biologists Limited, Cambridge, U.K.

regions of the exoskeleton to deposition of endocuticle which has helicoidal chitin orientation throughout its thickness. So it is possible to make locusts grow cuticle architecture to order; this could be of use in various chemical and physical investigations. Another use of this field is that the exact ages (counted from the daily layers) of many kinds of insects may be determined and applied in various behavioural, physiological, biochemical, and ecological contexts (Neville, 1983).

Experiments with leg epidermis and cuticle implanted into the blood haemocoel indicated that the circadian clocks involved were not located in either the nervous system or the endocrine system, but were probably in the epidermis itself (Neville, 1967b).

Cockroaches (*Periplaneta americana*) also have daily layers in the cuticle (Neville, 1963a). These are also caused by daily changes in chitin microfibril orientation. This has been confirmed in *Blaberus fuscus* (Lukat, 1978). Daily layers are still deposited after removal of the optic lobes; this causes the circadian locomotory rhythm to disappear. There must be at least two clocks in a cockroach. It had always seemed obvious to me that if a single-celled protist could have a circadian clock, then it was likely that each cell in a multicellular insect could also have one. I suggested that the epidermal cells in an insect are circadially rhythmic (Neville, 1975a).

Weber (1985) has successfully cultured leg integument from a cockroach (*Blaberus craniifer*) *in vitro*. Circadian-like layers were deposited, corresponding to the number of days in culture. Their appearance in a polarizing microscope was the same as that for control cockroaches. The material from this exciting result should now be analyzed by electron microscopy for confirmation of its ultra-structure (helicoidal in alternation with unidirectional?).

Unlike in locusts, the chitin ultra-structure of cockroaches cannot be changed during deposition by continuous light treatment. This has been tried in *Periplaneta americana* (Neville, 1967a) and in *Leucophaea maderae* and *Blaberus craniifer* (Wiedenmann, Lukat, & Weber, 1986). I was lucky to choose locusts for my initial experiments. Wiedenmann and associates conclude that the circadian mechanism driving the endocuticle rhythm in cockroaches will not become entrained by light.

Ougherram (1989) has suggested that the helicoidal layering in sclereid cells of olive (*Olea europea*) is circadian. It did not prove possible to measure the period, because of the complexity of cells growing out of phase with one another. But there seems no a priori reason to presume that 180° of helicoid should correspond to a twenty-four-hour period.

5.2i Epidermal light sense and chitin orientation

It was shown in Section 5.2h that the alternation of helicoidal and parallel layers of chitin in locust cuticle is normally coupled to a circadian clock; continuous light at constant temperature overrides this coupling. The site of this light reception has been shown not to be either compound eyes, ocelli, or living green foodplant (Neville, 1967b). Uncoupling still occurs in the endocuticle growing in a cylinder of leg integument implanted in the haemocoel. But it does not occur in an effectively painted intact leg. The response seems to be localized at the level of the epidermal cell and may be likened to a dermal light sense; see

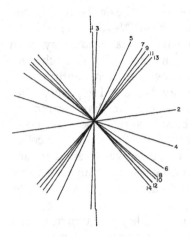

Figure 5.22. Main chitin fibril directions in beetle endocuticle. Plot of the fibre directions of the first fourteen unidirectional endocuticle layers (with parallel chitin crystallites) seen in surface view, at a chosen region on the hind coxa of an adult beetle (Dynastes cerebrosis; Coleoptera, Scarabaeidae). The clockwise rotation is seen clearly by considering separately the series of even- or odd-numbered layers. The fibril directions are accurate to ±3°. The plot resembles those for the fibril directions in fish scales (see Fig. 2.11), but the main layers in beetles are connected by thin intervening partial turns of helicoid. From Zelazny and Neville (1972), reprinted with permission of Pergamon Press.

the review by Steven (1963). On reflection, the term 'epidermal light sense' seems more appropriate for an insect.

5.2j Epidermal cell gradient and chitin orientation

Most experiments on the gradient which tells insect epidermal cells their position and polarity are analyzed by reference to cuticular surface structures (e.g. segments, hairs, scales, and bristles). We extended this to the interior ultra-structure of the cuticle by suggesting that control of specific directions of chitin microfibrils could also involve cell gradient and polarity (Neville & Luke, 1969b). This was demonstrated by Caveney (1973) using the pseudo-orthogonal layers of mealworm beetles (*Tenebrio molitor*). In the adult beetle the first layer of endocuticle has chitin microfibrils in parallel and arranged to follow the bilateral body symmetry. A square of larval integument may be transplanted to a different level in the segmental gradient, or rotated, say through 90°. The subsequent adult shows predictable disturbances to its chitin microfibrillar pattern. Further details are given by Caveney (1973) and Neville (1975a).

5.2k Accuracy of specific fibril directions

In pseudo-orthogonal systems (Section 1.3n) the thick layers are oriented in specific directions with respect to the various axes of the body. In the case of beetle endocuticle, these major layers are connected by helicoidal regions. Each helicoid turns left-handed through just less than 90°. This gives the impression that the major layers are turning clockwise, as seen through the light microscope (which cannot resolve the components of the intervening helicoids). Quantitative measurements and subsequent angular plots (Fig. 5.22) reveal a high degree of accuracy for each layer in the sequence. The layers may be deposited with an accuracy of ±3° (Zelazny & Neville, 1972). In another system which we now know to also be pseudo-orthogonal – the S-layers and intervening helicoids in wood tracheids – the S-layers are also laid down with great angular accuracy, in this case in the form of helices (Preston, 1974). This kind of accuracy would seem to demand directed assembly by the nearby cells; self-assembly on its own could not produce this result? We are still unable to offer an explanation for this degree of precision.

5.2l Are microvilli involved?

In insects, the shape and packing of the cellular microvilli which cover the outer surface of the epidermis are too irregular to explain the fibrillar architecture of the cuticle. However, in some animals, the microvilli, which lie perpendicularly to the cell sur-

5.2 Directed assembly: hypotheses / 175

Figure 5.23. Scindulene (shingle) hypothesis for collagen fibril orientation in a fish epidermal basement lamella. The fibrils are shown in white in surface view and are envisaged as emerging in overlapping groups from the cells which secrete them. The model was proposed by Nadol et al. (1969) to explain the apparent oblique crossing of microfibrils from one face of the basement membrane to the other. An alternative explanation, due to Bouligand (1972), is seen in Figure 5.24. Redrawn from Nadol et al. (1969).

face, may act as a template in the orientation of extracellular fibrils. For instance, Burke and Ross (1975) suggest that the orthogonal array of collagen fibrils in earthworm cuticle is organized by regularly arranged epidermal microvilli. However, Humphreys and Porter (1976) would reverse cause and effect, preferring to believe that the orthogonal plies of collagen fibrils form first, and that they then dictate the positions of the microvilli. That microvilli are not an absolute requirement for formation of orthogonal collagen structure is pointed out by Lepescheux (1988), who reasons that it is formed in the basement lamellae of many animals, on the inner surface of the epidermal cells; the plasma membrane on this surface does not form microvilli. In annelidan and vestimentiferan worm cuticles, Gaill (1990) considers that the packing symmetry of the microvilli determines the orientation of the collagen fibrils. The lattice is square in annelids and mainly hexagonal in vestimentiferans. In the latter the packing in the contractile part of the worm is orthogonal, but hexagonal in the non-contractile part. Gaill suggests that the third dimension acts as a stabilizer. Neville (1975a, p. 64) gives a diagram showing how square and hexagonal lattices of microvilli may interchange to produce a corresponding change in the orientation of chitin fibrils in the gut peritrophic membrane of cockroaches.

In the eggshell of the silk moth *Hyalophora cecropia*, isolated microvilli form a mould around which a cylindrical channel is made by the deposition of the surrounding eggshell proteins (Smith et al., 1971). This has been confirmed in *Bombyx mori* (Papanikolaou, Margaritis, & Hamodrakas, 1986).

It is concluded that in some cases a lattice of microvilli may act as a template, constraining fibrils to follow the lattice.

5.2m Shingles and screw-carpets

Two essentially similar models linking the cell surface directly with fibril orientation control were independently proposed at about the same time. One, the *shingle* or *scindulene* hypothesis, was proposed for collagen fibrils in the basement lamella of a fish (*Fundulus*) by Nadol et al. (1969). They envisaged that collagen fibrils do not lie parallel to the plane of either epidermis or basement lamella, but rather insert into both at a shallow angle, like tiles on a roof, and hence the name. Fibril direction would be determined by the direction of emergence from the epidermis – like the direction of cat fur – an analogy used by Gubb (1976). If we extend the analogy to a zebra and imagine that the fur on a white stripe points along the body axis, whereas that on a black stripe points across the body axis, then a surface view would show overlapping hairs (Fig. 5.23). The result would be a tilted orthogonal ply and could explain the apparent crossing of fibrils from one surface of the basement lamella to the other; this is

Figure 5.24. *A set of helicoidal layers in which the component fibres are curved (non-geodetic). This gives the false impression that ranks of arcs traverse obliquely across the laminate. This elegant explanation applies to some arthropod cuticles, as well as to fish basement lamella. In the latter, observers have been misled into thinking that layers of collagen fibres project obliquely from the epidermal cells (scindulene model of Nadol et al., 1969; see Fig. 5.23). Redrawn from Bouligand (1972).*

sometimes seen in electron micrographs. There is, however, an alternative explanation for this observation (Bouligand, 1972), in which a helicoid is built from curved microfibrils instead of straight ones – with the curves in the plane of the component layers (Fig. 5.24). (This should not be confused with arced patterning.) Oblique sections of such a non-geodetic helicoid could generate ranks of arced patterns which would appear (deceptively) to traverse obliquely between the two surfaces of the helicoid.

The other model, called *screw-carpet*, was proposed for chitin microfibrils in the cuticle of a locust (*Schistocerca*) by Weis-Fogh (1970) and was published in a book which I edited. He started from the supposition that synthesis and crystallization of chitin microfibrils were simultaneous events; this is fundamentally different from the approach of those of us who favour self-assembly explanations. Weis-Fogh's view was that the chitin microfibrils remained attached to the epidermal cell membrane, were secreted continuously at a shallow angle, and curved around in sheets, so that each sheet could contribute to more than a single level in a helicoid. His evidence was that after removal of resilin by enzymatic hydrolysis from locust pre-alar arm ligaments (which contain only chitin and resilin), this left behind chitin in the form of a continuous spring. He once showed me that such a spring could be extended or contracted like a concertina.

The screw-carpet model therefore resembles the shingle model. Which will prove to be correct in the long term – either rapid conventional crystallization or delayed self-assembly via a liquid crystalline phase – remains to be seen. I believe the self-assembly alternative to be the more productive, because both synthetic and natural model systems are available.

5.3 Mechanical reorientation

By contrast with Sections 5.1 and 5.2, which dealt with primary orientation, this section is about secondary reorientation. It concerns the changes in microfibrillar orientation brought about by physiological forces such as growth, muscular contraction, body weight, blood pressure, turgor pressure, and swelling by food intake. Some of the examples originated as models for primary orientation, but rightly belong in this section. One of the key problems is to disentangle primary from secondary fibre orientation (Neville, 1967c), as they sometimes coincide. For example, in plant primary cell walls, growth in dimensions and self-assembly of helicoids are happening at the same time.

5.3a Cellulose reorientation in plant cell walls

The classical *multinet growth hypothesis* (Roelofsen & Houwink, 1953) proposed that new cellulose fibrils are laid down more

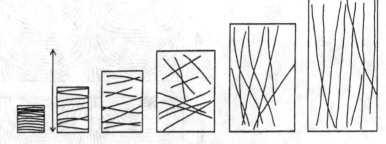

Figure 5.25. The classical multinet hypothesis of cellulose microfibril orientation in plant cell walls, as proposed by Roelofsen and Houwink (1953). The cell longitudinal axis runs north to south (arrow). An original square of wall (diagram on far left) is assumed to have its cellulose fibrils more or less perpendicular to the cell long axis. As it grows, it extends progressively in area – predominantly in the direction of the cell long axis. As the cell lengthens (diagrams in sequence from left to right), the cellulose fibrils reorient as shown. In many primary walls, however, the situation is now believed to be more complicated than this (see Fig. 5.26). Redrawn from Roelofsen and Houwink (1953).

or less perpendicularly to the cell longitudinal axis, becoming reoriented by the strain of turgor pressure during growth, so as to lie eventually more or less parallel to it (Fig. 5.25). Erickson (1982) computed the effect of the multinet hypothesis as it would appear in surface view. The hypothesis was renamed 'passive reorientation' by Preston (1982), who provided a formula for calculating the change in position of each microfibril due to growth strain; allowance was made for the original position of a microfibril. The earliest deposited microfibrils on the outer side of a cell wall are subjected to the greatest reorientation, and the most recent ones on the inner side of the wall to the least change in position. Preston (1982) inadvertently derived a formula which could be applied to primary walls which are helicoidal, although he assumed an initially random distribution of microfibrils. We were thus able to adapt Preston's formula to a cell which was laying down helicoidal wall and to predict by computer model what would happen to the microfibrillar patterns seen in section (Neville & Levy, 1984). Typical patterns derived from growth reorientation of helicoids, with various angles of rotation between adjoining layers, are given in Figure 5.26.

Our suggestions thus owed much to the thinking of previous workers, but the results were different because we considered helicoids as the starting point. We adopted various principles from the existing literature: (1) Microfibrils are added to the inner surface of the cell wall. (2) Microfibrils can be passively reoriented. This is likely to occur more extensively if the matrix remains stiff. If, however, the matrix becomes more extensible, by pH change induced by auxin (Fig. 5.12A), then less reorientation of microfibrils is likely to occur. This is because the microfibrils can slide through a plasticized matrix rather than rotate to change their orientations. We were surprised that the helicoidal cell walls of the stonewort *Nitella* showed no apparent change in architecture, despite their very great elongation during growth (Levy & Neville, 1986). Our computer models (Fig. 5.27) predicted noticeable change in pattern (Neville & Levy, 1984). Perhaps the cell walls of *Nitella* remain plasticized during growth, so that their microfibrils slide through the matrix to preserve

178 / *How is fibre orientation controlled?*

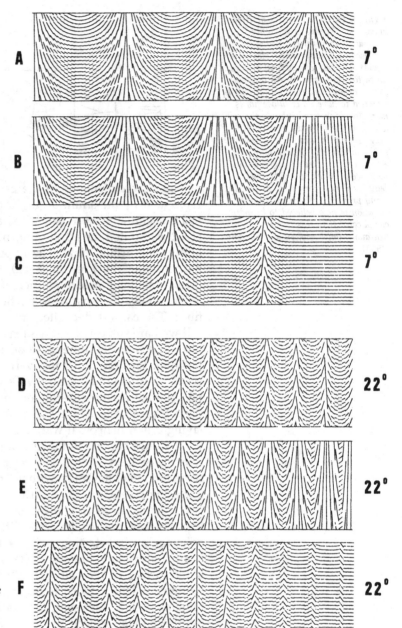

Figure 5.26. Computer models of plant cell walls. Computer graphics plots of predicted patterns of cellulose microfibrils in sections of strained helicoidal plant cell walls. The strain axis is vertical. The inner side of the cell wall (youngest, and therefore least strained) is on the left of each section. The outer side (oldest and most strained) is on the right of each section. (A–C) Simulated sections of helicoid with an initial rotation angle of 7° between component planes of microfibrils, such as occurs in Boergesenia (Mizuta & Wada, 1982). (A) Unstrained. (B) Strained and sectioned parallel to the strain axis. (C) Strained and cut perpendicularly to the strain axis. (D–F) Simulated sections of helicoid with an initial rotation angle of 22°, such as occurs in Cymodocea (Doohan & Newcomb, 1976). (D) Unstrained. (E) Strained and cut parallel to the strain axis. (F) Strained and cut perpendicularly to the strain axis. From Neville and Levy (1985), with permission of Cambridge University Press.

helicoidal architecture? Subsequent work on *Nitella* (Levy, 1991) has shown that helicoidal wall is deposited both before and after the active cell lengthening phase (during which microfibrils are deposited transversely to the axis of elongation). (3) A strain gradient acts across the wall thickness (Probine & Preston, 1961). In our computer plots (Figs. 5.26 and 5.27), this gives rise via a herring-bone type of pattern (Neville & Levy, 1984) to a thinning of helicoidal pitch, and then to a dissipated appearance in the outermost wall layers (Levy & Neville, 1986). The growth-induced dissipated appearance and thinning of layers were recognized in electron micrographs by Roland et al. (1982) in mung beans (*Vigna*). Micrographs of strained helicoidal systems with similar appearance are found in the literature, such as *Pisum sativum* (Appelbaum & Burg, 1971; Lang, Eisinger, & Green, 1982), *Pithophora oedogonia* (Pearlmütter & Lembi, 1978), *Pelvetia fastigiata* (Peng & Jaffe, 1976), *Cymodocea rotundata* (Doohan & Newcomb, 1976), and *Zea mays* (Satiat-Jeunemaître, 1981). Also, in the alga *Cylindrocapsa*, the parent cell walls remain after cell division, building up a set of shells like Russian dolls (Hoffman & Hofmann, 1975). These show varied distortion, with the outermost wall thinnest and most dissipated, and the youngest wall with the clearest arced patterning.

5.3b Chitin reorientation in insect cuticles

In blood-feeding bugs like *Rhodnius* the cuticle has to be able to expand at the appropriate time to allow ingestion of the huge volume of food. (Blood is extremely bulky, as it is mostly water.) Nuñez (1963) showed by cutting nerves to the epidermis on one side of the insect that expansion then occurred on the intact side only. Because cuticular expansion was accompanied by chitin microfibril reorientation, this experiment produced individuals with different chitin orientations on left and right sides of the body. Cuticular expansion involves plasticization of the cuticle matrix (Reynolds, 1973), probably by hydrogen ion secretion triggered by some neurohormone resembling 5-hydroxytryptamine (Fig. 5.12B). Reynolds showed that there was more extensive secondary reorientation of chitin microfibrils in unplasticized cuticle than in plasticized cuticle. This is probably because – as we just saw for plant cell walls (Section 5.3a) – micro-fibrils are able to slide within the plasticized matrix. The force driving expansion, and hence chitin reorientation in *Rhodnius*, is muscular.

In the more highly evolved dipteran flies, the cuticle of the last larval instar shrinks along its long axis, and expands around its circumference, to form a barrel-shaped puparium (the true pupa is contained within the puparium). These forces reorient the originally helicoidal chitin so that its preferred orientation is around the circumference of the puparium. At emergence of the adult fly, an inflatable structure (the ptilinum) is then able to break open a circular cap at the end of the puparium, assisted by the cleavage provided by the new (parallel) chitin orientation. The reorienting force is now known not to be muscular in origin (Zdarek & Fraenkel, 1972).

We have shown by laser optical diffraction of the cuticle of an adult beetle (*Oryctes rhinoceros*) that the packing of chitin

Figure 5.27. Computer-generated model of the expected appearance of an electron micrograph of a plant cell wall. The predicted pattern is of an oblique section through a helicoidal cell wall which is elongating. The axis of elongation runs parallel to the horizontal, and the upper layers represent the first to be deposited. These outermost layers would be expected to contain the most reoriented cellulose microfibrils and also to be subject to the maximum thinning of layers as the cell elongates. The original rotation angle is 16°. The arcs due to the helicoid are distorted firstly to herring-bone and then to a dispersed appearance. From Levy and Neville (1986), by permission of Dr. B. Vian, Paris.

crystallites changes during expansion at the moult from hexagonal to distorted hexagonal (Neville, Parry, & Woodhead-Galloway, 1976). In a little-known paper using X-ray diffraction on insect cuticle, Kroon, Veerkamp, and Loeven (1952) showed that during expansion of the wings in a small tortoiseshell butterfly (*Aglais urticae*), the chitin crystallites in the wings became secondarily reoriented, mainly along the vein axes.

5.3c Examples where mechanical orientation is not involved

Dennell (1976) proposed that chitin orientation may be determined by mechanical stresses related to body form. We know of several examples where this is not so: (1) Arced patterning has been seen in electron micrographs of cuticle grown *in vitro* by imaginal disc cells of *Plodia interpunktella* (Dutkowski, Oberlander, & Leach, 1977). (Imaginal discs, found inside the thorax of larvae and pupae, develop into the wings of endopterygote insects.) This shows that helicoidal cuticle can be formed in the absence of mechanical stresses. (2) The asymmetrical nature of the helicoidal chitin in insect cuticle, and of collagen in bird cornea, together with all other helicoidal systems in bilaterally symmetrical animals, shows that its orientation could not have a mechanical origin. (3) Five-day-old chicks (*Gallus domesticus*) are still capable of depositing twisted orthogonal collagen architecture, even after puncture of the retina, thus lowering the pressure in the ocular fluid (Bard, 1990). (4) In Section 5.2h it was shown that the chitin orientation of locusts can be changed by altered light regimes, even though the shape of the locust remains constant. This could not happen if the chitin orientation was under mechanical control. (5) If mechanical forces determine chitin microfibril directions, then those of a locust leg would be expected to lie at 45° to its axis, but they do not (Neville, 1967c). (6) Implanted cylinders of locust leg integument lay down chitin microfibrils with the same orientations as in a normal leg (Neville, 1967b). Because no skeletal forces act on a cylinder floating freely in the haemocoel, we can deduce that mechanical forces do not create the architecture of locust cuticular fibrils.

In summary, self-assembly of molecules seems adequate to explain the formation of simple systems such as helicoids. When specific fibre orientations are concerned, directed assembly may be required, but we cannot yet explain it. In some situations both self-assembly and directed assembly may act in concert, and both may be altered by secondary reorientation.

6

Unifying themes

6.1 Interrelations and transitions between different architectures

Different types of fibrous composite architectures are often seen together, sometimes with one type grading into another. By using several examples of such transitions, nine types of structures have been linked together by arrows in a diagram (Fig. 6.1). This diagram is an illustrated extension of the one constructed by Bouligand and Giraud-Guille (1985). This approach suggests a common origin for such structures and also integrates liquid crystalline analogues. If a mobile liquid crystalline phase proves eventually to be involved in the development of these systems, it will show how it is possible to change architecture, either *from location to location* or *in sequence*. In either case, different types of structures may *co-exist*.

Detailed evidence for the *linkage analysis* used in constructing Figure 6.1 is given later. Use is made of information from both plants and animals. The nine types of architecture represent variations on a few themes; these are parallel, orthogonal (which may be regarded as a 90° helicoid), and various other angles of helicoids (e.g. 60°, 45°, 10°, 1°). Other orientations are random, planar, polydomain, and cylindrical. The purpose of constructing Figure 6.1 is to show common principles in structure and development across a range of plant and animal skeletons.

The *transitions* between structural types are thought to rely upon the transient *mobility* of the fibrous component within the matrix. Liquid crystals are mobile and show transitions between textures. Figure 6.2 shows that liquid crystal phases may interchange from parallel nematic to twisted nematic, which may roll up to form a cylinder, a group of which can in turn form a blue phase. These phase changes constitute the basis for the transitions in Figure 6.1.

The transition of different architectures may be illustrated by a specific example. When heated, different mixtures of the compounds shown in Figure 6.3 co-polymerize to preserve in the solid state the liquid crystalline order of nematic, smectic, or cholesteric textures. These examples have been used by Bouligand and Giraud-Guille (1985) as analogues of liquid crystals.

The fact that helicoidal and parallel architectures may be experimentally interchanged in developing locust cuticles argues

182 / *Unifying themes*

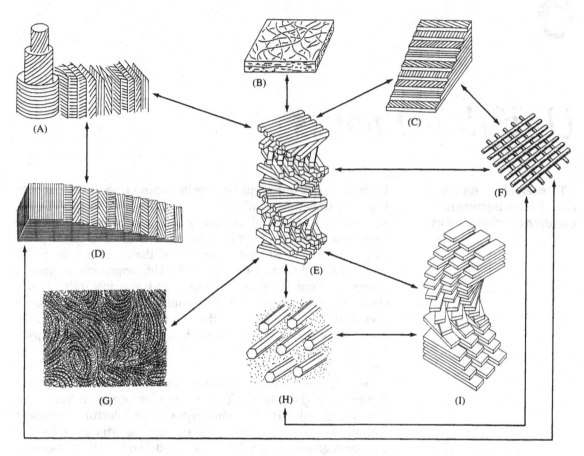

Figure 6.1. Evidence for a common origin of the diversity of fibrous composite architectures. (A) Cylindrical helicoidal *(double twist)* grading into planar helicoidal *(single twist)*. Example: bone haversian system. This is the analogue equivalent of blue-phase liquid crystals. Bone can change from helicoidal to orthogonal within an individual cylindrical osteon. (B) Planar random layer. Examples: parts of some plant cell walls. (C) 45° helicoid. Example: dogfish eggcase (Scyliorhinus) (Knight & Hunt, 1974b). Other systems are arranged at other angles (e.g. 60° in Polynoe scale worm) (Bassot, 1966). (D) Twisted orthogonal. Examples: fish scales, vertebrate cornea (fish, frog, turtle, birds). (E) Monodomain helicoidal *(small rotation angle)*. Examples: insect cuticles, some plant cell walls, many cysts and eggshells. This is the analogue equivalent of cholesteric liquid crystals. (F) Orthogonal. Examples: basement lamella in vertebrates and a flatworm; cuticles of cylindrical animals. (G) Polydomain helicoid. Example: mantis eggcase proteins. (H) Parallel *(unidirectional)*. Examples: tendons in arthropods and vertebrates, daytime layers in insect cuticle. This is the analogue equivalent of nematic liquid crystals. (I) Pseudo-orthogonal. Examples: wood tracheids, endocuticle in some parts of adult beetles, bugs, stick insects, and dragonflies. Layers only appear to change abruptly through a large rotation angle; there is actually an intervening thin sandwich of helicoid.

in support of their interrelation as part of a transitional series. Other experimentally induced transitions include a change of helicoidal pitch in colchicine- or carbamate-treated plant cell walls (Section 5.2d) and the change from twisted orthogonal to orthogonal structure in DON-treated bird cornea (Section 5.1d).

The sources of evidence which justify the arrowed linkages between the structures in Figure 6.1 are given below, with the

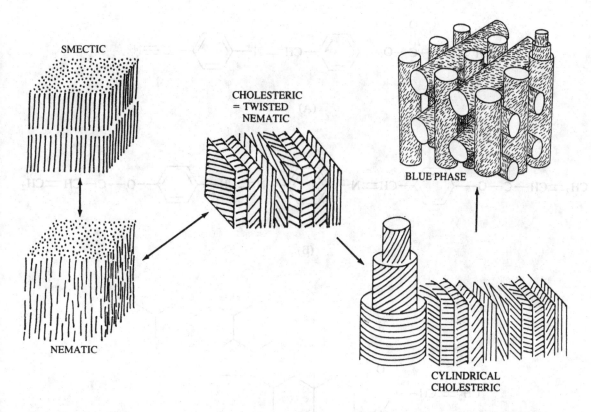

Figure 6.2. Diagrams to show how the structures of five types of liquid crystals may undergo transition from one sort to another. Compare with the types of fibrous architectures in Figure 6.1. It is feasible that these derive from liquid crystals which have become stabilized, by crystallization to the solid state.

letters classified in alphabetical order. Helicoids clearly occupy a pivotal position.

AD (cylindrical helicoidal–twisted orthogonal): bone osteons.
AE (cylindrical helicoidal–monodomain helicoidal): compact human thigh bone.
BE (planar random–monodomain helicoidal): some plant cell walls.
CE (45° or 60° helicoid–monodomain helicoid): mantis eggcase proteins.
CF (45° or 60° helicoid–orthogonal): mantis oothecal proteins.
DF (twisted orthogonal–orthogonal): bird cornea. Twist stops during development, or with treatment by DON.
EF (helicoidal–orthogonal): *Chamaedoris* (green alga) changes to helicoidal cell wall in response to APM treatment; *Oocystis* (alga) changes to helicoidal cell wall if treated with MBC; nematode cuticle (*Mermis nigrescens*); mantis eggcase proteins; chitin in periostracum of whelk (*Buccinum undatum*); defensive protein of sea cucumber (*Holothuria forskali*); sea-squirt tests

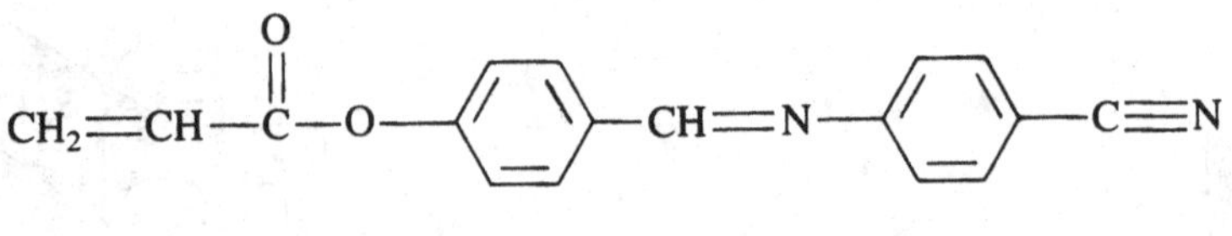

Figure 6.3. Some synthetic liquid crystal/solid crystal models. Some liquid crystals polymerize to the solid state when mixed and heated. Various combinations of substances A, B, and C can form smectic, nematic, or cholesteric solids: A, p-acryloyloxybenzylidene-p-cyanoaniline; B, di-p-acryloyloxybenzylidene-p-diaminobenzene; C, cholesterol acrylate. These synthetic models are important, as they demonstrate the existence of a range of architectures with interlinking transitions. Similar linkage is attempted for biological architectures in Figure 6.1.

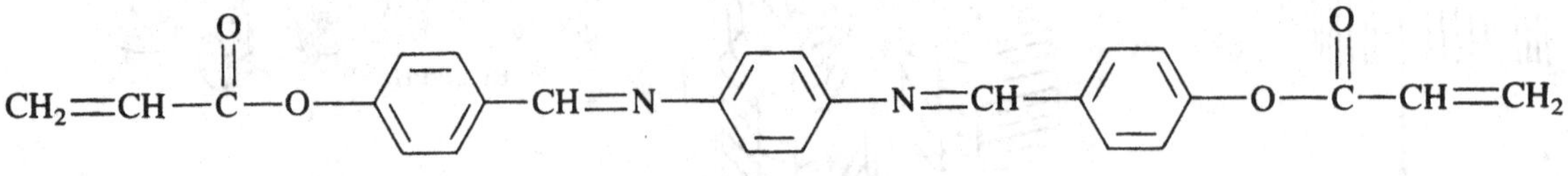

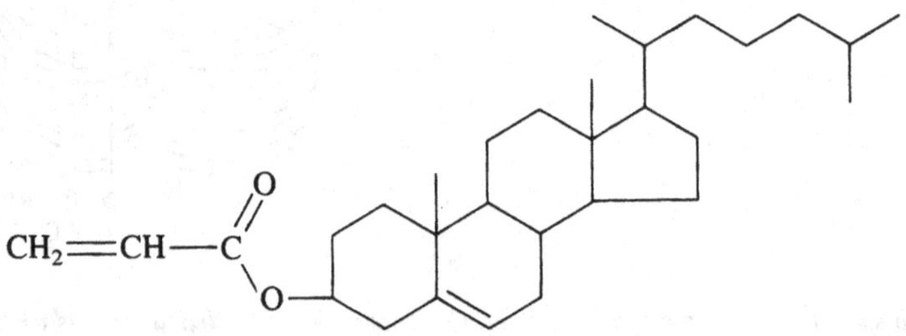

(*Halocynthia papillosa* has helicoidal test, but *Microcosmus* has orthogonal test); human bone osteons.

EG (monodomain helicoid–polydomain helicoid): mantis eggcase proteins.

EH (helicoidal–parallel): daily alternation in insect cuticle, changeable by light regime; quince seed epidermal mucilage; pseudo-orthogonal systems classified under I; sea cucumber defensive protein is helicoidal before and parallel after eversion.

EHI (helicoidal–parallel, and pseudo-orthogonal in different regions of the stick insect (*Heteropteryx dilatata*).

EI (helicoidal–pseudo-orthogonal): Transition at moult from exocuticle to endocuticle deposition in adult bugs, dragonflies, stick insects, and beetles.

HF (parallel–orthogonal): non-mammalian vertebrate corneas; parallel layer is component of orthogonal.

HI (parallel–pseudo-orthogonal): beetle endocuticle and wood; parallel layer is component of pseudo-orthogonal.

6.2 Some general principles of helicoids

This section is about helicoids, because we know more about them than other architectures. It should serve as a summary of present knowledge. The *energy* required to convert nematic into cholesteric liquid crystals is low. Helicoidal composites are not only abundant in living systems (Chapter 2) but also successful. Similar problems can be solved by similar textures, and yet be made from entirely different biopolymers. Helicoids are adaptable, *moulding* into virtually any shape.

6.2a Origin and stabilization of the liquid crystalline state

Composites serving skeletal functions are built with molecules which are *large*, nearly always *extracellular*, and *water-insoluble*. Control over their orientation requires that the continuous medium of the composite passes through a *mobile* stage. It is likely that *self-assembly* creates a liquid crystalline phase. The fluidity of liquid crystals varies with their three-dimensional order (Frank, 1958). When they have little order they are more mobile, but they become less mobile as order increases. The driving force behind such developing order is the need to achieve a low state of *free energy*.

Helicoidal self-assembly takes *time*; the assembly of mantis eggcase proteins between slide and cover-slip takes a day or two (Neville & Luke, 1971b). Three days were required for cholesteric self-assembly of collagen in the preparations of Bouligand et al. (1985); the time doubled in the presence of calf serum. Self-assembly of collagen fragments into the cholesteric phase by the method of Giraud-Guille (1989) took only a few hours.

To function as mechanical composites, liquid crystalline arrays must *consolidate* (see Section 4.2h). An example of this in a synthetic system is given by Bhadani and Gray (1984). The acrylic acid ester of hydroxypropyl cellulose forms a cholesteric phase which reflects interference colours with circularly polarized light. This indicates helicoidal structure (see Section 3.3b). Ultra-violet light treatment for five hours stabilized these liquid crystals. The colours and optical activity were preserved as a 'locked-in' cholesteric organization. The biological equivalents of this consolidation could include cross-linking of protein chains by enzyme-controlled free-radical molecules.

Bonding of microfibrils to matrix has been covered in Section 3.1c. An example from insect cuticle is given by Hillerton (1980); only some of the proteins were extractable, with the remainder forming a well-bonded sheath around each chitin microfibril. In plant material, the bonding of xylans onto the surfaces of cellulose microfibrils has been established by studies on wood pulp (Mora et al., 1986).

A recently detected unifying principle in plants and animals

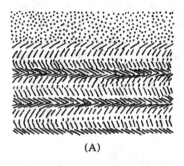

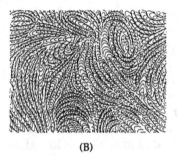

Figure 6.4. Planar monodomain versus polydomain helicoids. (A) Diagram of an oblique section through a planar monodomain helicoid, in which the component microfibrils lie in parallel planes. The helicoidal layers are secreted up against the surface of an initial restraining layer (stippled) which aligns the microfibrils into a planar and single domain. (B) As for part A, except that the helicoid is polydomain, lacks a constraining layer, and thus has no laminated alignment. The helicoid still gives rise to ranks of arced patterns, but these are not aligned with any initial constraining layer. Redrawn from Neville (1988b).

concerns the way in which the shapes of the liquid crystalline phases are moulded by their surroundings (Neville, 1988b). If a cholesteric liquid crystal is trapped between the surface of the cell which secreted it and a previously secreted layer, its shape is constrained, usually to either a planar monodomain helicoid, which is flat (Fig. 6.4A), or to a concentrically curved shell. A thorough search of the literature on helicoids in plants and animals revealed that in all cases a non-helicoidal layer preceded the deposition of helicoids (Neville, 1988b). Examples included the broad categories of plant cell walls, spores, animal eggshells, and cuticles. The initial layer serves to constrain the helicoidal layers so that they are moulded in parallel against it.

The shape of the periplasmic helicoids in quince seed epidermal cells has been reconstructed from serial transmission electron micrographs (Abeysekera & Willison, 1990). The liquid crystalline helicoids become moulded to the shape of the cell walls. We have seen that the crystallites in quince mucilage (*Chaenomeles japonica*) show a tendency to line up next to the cell wall surface (Martin & Neville, unpublished observations). Similar observations were made on cholesteric spherulites of self-assembled collagen (Giraud-Guille, 1989); planar domains were seen where the spherulites were constrained between glass slide and coverslip. Robinson (1966) made similar observations on cholesteric liquid crystals of PBLG, placed for a few days in a specially constructed container with optically parallel walls.

If cholesteric liquid crystals are not constrained by smooth surfaces, they adopt a much less regular form – known as helicoidal polydomain (Fig. 6.4B). Polydomain cholesteric liquid crystals are found in mantis oothecal proteins and sea cucumber defensive secretions. They may be experimentally induced in systems which are normally monodomain helicoids. Thus, in mung beans, epidermal cell walls may be caused to contain whorling patterns by application of an osmotic shock brought about by immersion in 0.5-M mannitol (Reis, Roland, & Vian, 1985). The cells return to making planar monodomain helicoidal walls on restoration of normal cell turgor pressure. Normal pressure is required for the deposition of monodomain helicoidal plant cell wall layers; the pressure helps to constrain newly secreted wall components between plasma membrane and cell wall. In an animal system, my polydomain explanation can be used to interpret the whorled patterning seen in repair cuticle in Crustacea. This irregular type of cuticle architecture is secreted after physical wounding of the integument in the crab *Carcinus maenas* (Halcrow & Smith, 1986).

Arthropod cuticle is the result of the *pooled* secretions of the epidermal cells. The continuity of arced patterning (Fig. 6.5A) indicates that the helicoidal secretions of neighbouring cells join up neatly. This is circumstantial evidence for cuticle self-assembly. If the secretions did not self-assemble, the result would be a

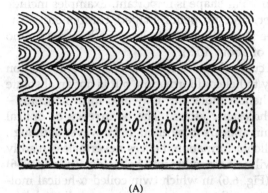
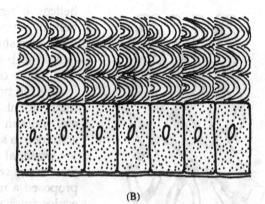

Figure 6.5. (A) Diagram of the arced fibrillar pattern seen adjoining the epidermis in an oblique section of helicoidal arthropod cuticle. The continuity of pattern indicates that the products of neighbouring cells join up neatly; this is indirect evidence for cuticle self-assembly. Otherwise there would be a series of faults opposite the cell boundaries, as in part B. This would be mechanically weakening, and is not observed.

mechanically unsound series of fault planes opposite the cell boundaries, as in Figure 6.5B; this does not occur. Even when the secretions of individual cells are not pooled, as in plant cell wall formation, the timing of deposition of layers by neighbouring cells may still be in phase (Mosiniak-Bessoles, 1987).

6.2b *The twist is chiral*

With very few exceptions, natural helicoids always twist with a single sense of rotation for each type of material (Section 5.1e). They are therefore asymmetrical with respect to the whole body. This is particularly noticeable in bilaterally symmetrical animals, and it led me to write a simple book (Neville, 1976). It provides circumstantial evidence in support of origin of helicoids by self-assembly. It is likely that the singular twist of biological helicoids is executed by the same molecular asymmetry which gives rise to their optical activity – namely, construction from D-sugars and L-amino acids. While left/right helicoid asymmetry is determined by molecular *isomerism*, dorso/ventral symmetry is influenced by gravity, and antero/posterior symmetry by locomotion and cephalization. Functionally, it is preferable for skeletal materials to have one sense of rotation, as the components then fit together without zones of mechanical weakness or optical confusion.

6.2c *Molecular shapes: helices and grooves, pipe-cleaners, and propellors*

At the molecular level, there are several different ways to generate helicoids by self-assembly. A variety of chemicals can form

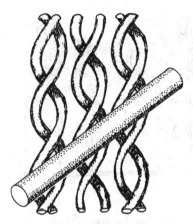

Figure 6.6. Possible relation between molecular conformation and angular rotation in mantis eggcase proteins. Mantis oothecins are thought to be coiled as twin α-helices. If these are packed in a flat sheet with a slight stagger (out of register), they create a system of lined-up grooves. The next sheet of twin α-helices may then lie in the grooves, as shown diagrammatically by the overlying cylinder. This could give rise to an angular rotation which, if repeated, would generate a helicoid. Adapted from Rudall (1956).

helicoids; hence molecular shape is important. Examples include helical, pipe-cleaner, and propellor-shaped components. The molecules should be straight and quite long, with an axial ratio (length to breadth) of about 10 : 1.

In terms of molecular shape, *helical* molecular conformation lends itself naturally to helicoid formation, but is not an absolute requirement. (Non-helical molecules may also form helicoids – e.g. xanthan and cholesteric esters.) The occurrence of optical activity in a stiff chain appears to be enough. However, although not essential for helicoid formation, helices nevertheless readily form ropes which will self-assemble into helicoids. Rudall (1956) proposed a model (Fig. 6.6) in which twin-coiled α-helical molecules from praying mantis eggcase protein were aligned in a sheet with a stagger; this lines up the helical grooves into which will fit the adjoining sheet of α-helices. Rudall's proposal was for just two sheets of helices, but if extended it could generate a helicoid. Bouligand (1978b) reasons that any subsequent sheet of helices would have a choice of lining up in parallel either with the grooves between twin α-helices or with the spiral grooves created by the stagger (Fig. 6.6).

Bouligand also noted that Rudall's model fits the helicoidal assembly of α-helices of PBLG; the benzyl groups stack helically to produce oblique grooves. He also points out that the Rudall model (which has a fixed angle) does not explain helicoids with variable pitch.

The building unit of collagen is the triple helix, assembled from three molecular chains (Fig. 4.13A,B). Three left-handed helical chains form a right-handed triple helix, a group of which may form a left-handed helicoid. (The changes in sense at each level in the hierarchy give tighter molecular packing.) We know that *in vitro* collagen can form a helicoid (Giraud-Guille, 1989), even in the absence of glycosaminoglycan matrix. (*In vivo* the matrix would be expected to play a part.) It is therefore valid to consider the molecular shape of collagen with respect to helicoid assembly; a triple helix seems well suited. A diagram showing the alignment of triple helices to form the basis of a helicoid is given in Figure 6.7.

The credentials of *pipe-cleaner*-shaped molecules which form helicoids – namely, synthetic cellulose derivatives (Gray, 1983) and hemicelluloses (Neville, 1988a) – are a stiff backbone chain, bulky flexible sidechains to allow the backbone to move as if it were in a liquid crystal, and an asymmetric carbon in each backbone residue. The longer the sidechains, the weaker are the main chain interactions, and hence the longer the pitch of the helicoid. An example of a pipe-cleaner molecule is xanthan, which is an extracellular secretion of a bacterium (*Xanthomonas campestris*) that can form cholesteric liquid crystals (Maret, Milas, & Rinaudo, 1981). Xanthan has a cellulose backbone with trisaccharide

Figure 6.7. *Possible relation between molecular conformation and angular rotation in collagen helicoids. Collagen chains entwine to form triple helices (see Fig. 4.13). If these pack in a sheet with a stagger (out of register), they create a system of lined-up grooves. The next sheet of triple helices may then adopt either of two possible stable positions. These are (A) in parallel with the initial layer in the grooves between triple helices or (B) in the grooves lying at an angle across the initial layer. The position shown in part B could give the angular rotation required to generate a helicoid. The sense of twist changes from triple helix to helicoid, because the different levels in the hierarchy then pack together in a more stable manner. (This principle is found in rope.) Based on the interpretation of sponge collagen (Garrone et al., 1973) and on diagrams by Gaill and Bouligand (1985) and Bouligand and Giraud-Guille (1985).*

(A)

(B)

sidechains of β-D-mannopyranosyl-(1 → 4)-β-D-glucuronic acid-(1 → 2)-α-D-mannopyranoside-6-O-acetate. These sidechains are attached at C-3 of alternate glucose residues of the main chain (Dentini et al., 1984).

Another molecular shape (propellor shape) which can form helicoids is the twisted β-pleated sheet (Hamodrakas, 1984). In the case of moth eggshell proteins there is a central conservative core with two flexible arms. Neighbouring molecules bond together, their cores forming fibrous crystallites, and their arms forming the matrix. The two components of the fibrous composite are bonded together by peptide bonds as in all protein backbones.

Although no systematic analysis has been made, I have formed the impression that the diameter of helicoidal construction units relates to the helicoidal angle. Thus, the large-diameter macrofibrils in sea-squirt test (*Halocynthia*) and in crab cuticle (*Carcinus*) are set at large angles to each other. The medium-diameter microfibrils, such as the chitin crystallites in insect cuticle, are set at a medium angle, whereas the individual molecules (α-helices) of liquid crystalline PBLG are set at much smaller angles.

6.2d Liquid crystals and fibrous composites: an intellectual gap?

In classical liquid crystals, interaction takes place between individual molecular chains. By contrast, in many of the helicoids of living systems, interaction takes place between individual molecules and microfibrils – forming a fibrous composite. This tends to create an intellectual gap, making it hard for some physical chemists to accept a liquid crystalline origin for some of the biological examples. The successful self-assembly into a helicoidal structure of a mixture of cellulose *crystallites* and glucuronoxylan *molecules* (both extracted from quince seed epidermal mucilage) is therefore an important result (Reis et al., 1991).

6.2e Control of helicoidal pitch

Examples which show that the pitch of helicoidal materials is controlled by the cells which secrete them have been given in Section 4.2c. A good description of a possible control mechanism (in the context of quince mucilage) is given by Abeysekera (1989). She suggests that controlled secretion of the concentrations of hemicellulose matrix and cellulose microfibrils could account for variation in helicoidal pitch. The hemicellulose, acting as helper molecule, could control cellulose orientation. Her suggestion is based upon the properties of some mixtures of synthetic compounds. In a multi-component system, 'effective rotary power' depends on relative concentration and varies inversely with helicoid pitch. Components with small pitch have large angular twist between consecutive layers. For example, in a mixture of oleyl cholesteryl carbonate with anisylidene p'-n'-butyl-aniline (ABUTA), helicoidal pitch increases (and the angular twist decreases) as the proportion of ABUTA increases. Again, in mixtures of cholesteryl nonanoate and cholesteryl chloride, the sense of helicoid can be reversed by changing relative concentrations. Cholesteryl iodide can be made to rotate right-handed by mixture with cholesteryl chloride, and left-handed when mixed with cholesteryl hexanoate.

6.2f Beyond the cell membrane

The topics in this chapter have so far concerned the development of fibrous composites in the extracellular environment. But what of the cells, which, so far as fibrous composites are concerned, appear to play an *indirect* role?

One situation where the cells appear to play a more *direct* part is in the orientation of the initial layer of a helicoid. It was shown in Section 6.2a how the formation of a non-helicoidal layer provides a constraint against which a cholesteric liquid crystalline phase can become aligned. Subsequently, when a helicoid has started to form, the orientation of the latest helicoid layer determines the direction of the next. But how is the direction of the very first helicoidal layer established? The simple answer is that we don't know.

In the egg cell walls of the seaweed *Pelvetia fastigiata*, strings of cellulose synthetase particles are oriented on the cell membrane in the same direction as the first layer of microfibrils of the wall helicoid (Peng & Jaffe, 1976). The particles are, however, aligned only for the first four hours of wall deposition, after which they are oriented at random. The directions of the initial layer of collagen fibres in the cornea of both eyes of a bird (*Gallus*) obey bilateral symmetry (Coulombre & Coulombre, 1975). Likewise with the first layer of endocuticle in an adult beetle, *Dynastes* (Zelazny & Neville, 1972). These observations imply that the cells have intimate control over these initial layer directions, even though the symmetry is later disobeyed in subsequent layers.

The cells must supply their immediate external environment (e.g. periplasm, deposition zone) with components in the correct mixture, at the correct times. In the case of collagen, short lengths of triple helix may be able to self-assemble, but longer pieces may need to be oriented by cell-directed assembly.

In general, whereas self-assembly may occur at quite remote distances from the cells, directed assembly requires intimate cellular control. The region beyond the cell membrane is just as important as the cellular compartments and the plasma membrane itself.

6.3 Evolutionary aspects

Although they are not universally recognized, most biologists would accept that living organisms may be grouped into *five kingdoms*. Helicoidal structures are found in all of them, implying that they appeared early in evolutionary terms. For details, see the tables in Chapter 2.

(1) *Prokaryotes* (bacteria and cyanobacteria). As noted in Section 6.2c, the bacterium *Xanthomonas* secretes xanthan, an extracellular polysaccharide which forms helicoidal liquid crystals. Helicoids also occur as a mucilaginous secretion of some blue-green algae (cyanobacteria), such as *Rivularia* (Section 4.3g). These observations suggest that helicoids have been around for a very long time.

(2) *Protists* (small algae and protozoans). Helicoidal structure is found in the cyst walls of some forms (see Chapter 2); so single cells are competent to make helicoids.

(3) *Plants*. Helicoidal cell walls are found in all of the major groups which have been investigated – cycads still require study.

(4) *Fungi*. So far, helicoids have been found only in spore walls of some primitive Zygomycetes. There are as yet no indications of helicoidal walls in slime moulds, Ascomycetes, or Basidiomycetes.

(5) *Animals* (Metazoa). Helicoids have been found in all those major phyla which have been investigated. They occur in a variety of products (cuticles, basement lamellae, corneas, bones,

tests, eggshells, and cysts). We have found helicoids in the cuticle of a fossil lobster from the Jurassic period.

Fibrous composites in general, and helicoids in particular, seem to have evolved convergently on many occasions. This may be deduced from the variety of biopolymers which are found in both fibre and matrix components. Plants and animals, especially terrestrial ones, have often evolved similar solutions when faced with the same ecological problems. Efficient support, which demands strength as well as lightness, is one such problem.

I have speculated elsewhere about the evolution of helicoidal walls in plants (Neville, 1988c). The walls of primitive (aquatic) algae probably were not fibrous composites; they would have had flexible and labile walls made of self-assembling glycoproteins – like those of *Chlamydomonas*. Such walls permit genome-capture of prokaryotes, leading to eukaryotes by endosymbiosis. Thus, a prokaryote system capable of producing and secreting helicoids – like *Rivularia*, *Chroococcus*, or *Xanthomonas* – could have been conveyed to higher forms. Only with the evolution of fibrous composite walls could plants have become multicellular, large, and terrestrial. The supporting tissues of higher plants have helicoidal walls (e.g. epidermis, collenchyma, sclerenchyma fibres and sclereids, and tracheids).

Plant cell walls contain branched hemicelluloses of complex shape (Section 3.1a). These have been shown to be the source of branched oligosaccharide hormones (Albersheim & Darvill, 1985). These are released from the wall hemicelluloses by enzymes, themselves activated by hydrogen ions released by auxin acting on the cell plasma membrane (Rayle, 1973). The small branched oligosaccharins typically contain nine residues and work in very low concentrations to control such plant activities as growth and flowering. I have proposed (Neville, 1988a) that complex hemicelluloses may have evolved originally in adaptation to forming helicoidal walls; this complexity then made possible (after the development of suitable enzymes) the release of oligosaccharins with the specificity of molecular shape demanded of hormones.

In the case of animals, the evolution of extracellular basement lamella made possible the coordination of individual cells into tissues. Sponges, which lack basement lamella, are unable to function at the tissue level. Insect epidermal cells sit, by contrast, upon a basement lamella and produce pooled secretions from their apical surfaces. Chordates show the further sophistication of cells positioned *within* the skeletal materials themselves (e.g. tests, bone, and cartilage).

Extracellular composites are deceptive; they play a more dynamic part in developmental biology than is generally appreciated. It is hoped that this book will encourage wider interest in the biopolymers involved.

References

Abeysekera, R. M. (1989). Helicoidal order in the periplasmic cellulose arrays of quince seed epidermis; a hypothesis for its origin. In *Cellulose and Wood Chemistry and Technology*, ed. C. Schverch, pp. 783–94. New York: Wiley.

Abeysekera, R. M., & Willison, J. H. M. (1987). A spiral helicoid in a plant cell wall. *Cell Biology International Reports*, 11, 75–9.

Abeysekera, R. M., & Willison, J. H. M. (1988). Development of helicoidal texture in the prerelease mucilage of quince (*Cydonia oblonga*) seed epidermis. *Canadian Journal of Botany*, 66, 460–7.

Abeysekera, R. M., & Willison, J. H. M. (1990). Architecture of the fluid cellulosic arrays in the epidermis of the quince seed. *Biology of the Cell*, 68, 251–7.

Adair, W. S., & Mecham, R. P., eds. (1990). *Organization and Assembly of Plant and Animal Extracellular Matrix*. New York: Academic Press.

Ahnelt, P. (1984). Chaetognatha. In *Biology of the Integument. I. Invertebrates*, ed. J. Bereiter-Hahn, A. G. Matoltsy, & K. S. Richards, pp. 746–55. Berlin: Springer.

Albersheim, P. (1975). The walls of growing plant cells. *Scientific American*, 232 (4), 80–95.

Albersheim, P., & Darvill, A. G. (1985). Oligosaccharins. *Scientific American*, 254 (3), 44–50.

Alberts, B., Bray, D., Lewis, J., Raff, M., Roberts, K., & Watson, J. D. (1989). *Molecular Biology of the Cell*, 2nd ed. New York: Garland Press.

Alexander, R. McN. (1987). Bending of cylindrical animals with helical fibres in their skin or cuticle. *Journal of Theoretical Biology*, 124, 97–110.

Alexander, R. McN. (1988). *Elastic Mechanisms in Animal Movement*. Cambridge University Press.

Andersen, S. O., & Hojrup, P. (1987). Extractable proteins from abdominal cuticle of sexually mature locusts, *Locusta migratoria*. *Insect Biochemistry*, 17, 45–51.

Andersen, S. O., Hojrup, P., & Roepstorff, P. (1986). Characterization of cuticular proteins from the migratory locust, *Locusta migratoria*. *Insect Biochemistry*, 16, 441–7.

Andersen, S. O., & Weis-Fogh, T. (1964). Resilin, a rubberlike protein in arthropod cuticle. In *Advances in Insect Physiology, Vol. 2*, ed. J. W. L. Beament, J. E. Treherne, & V. B. Wigglesworth, pp. 1–65. London: Academic Press.

Anderson, E. (1967). The formation of the primary envelope during the oocyte differentiation in teleosts. *Journal of Cell Biology*, 35, 193–212.

Anderson, E., & Huebner, E. (1968). Development of the oocyte and its accessory cells of the polychaete *Diopatra cuprea* (Bosc.). *Journal of Morphology*, 126, 163–98.

Anseth, A. (1965). Corneal wound healing. In *Structure and Function of Connective and Skeletal Tissue*, ed. S. Fitton-Jackson, R. D. Harkness, S. M. Partridge, & G. R. Tristram, pp. 506–7. London: Butterworth.

Appelbaum, A., & Burg, S. P. (1971). Altered cell microfibrillar orientation in ethylene-treated *Pisum sativum* stems. *Plant Physiology (Bethesda)*, 48, 648–52.

Arnott, S., Rees, D. A., & Morris, E. R., eds. (1983). *Molecular Biophysics of the Extracellular Matrix*. Clifton, N.J.: Humana Press.

Arsenault, A. L., Castell, J. D., & Ottensmayer, F. P. (1984). The dynamics of exoskeletal-epidermal structure during moult in juvenile lobster by electron microscopy and electron spectroscopic imaging. *Tissue and Cell*, 16, 93–106.

Bailey, A. J., Gathercole, L. J., Dlugosz, J., Keller, A., & Voyle, C. A. (1982). Proposed resolution of the paradox of extensive crosslinking and low tensile strength of Cuvierian tubule collagen from the sea cucumber *Holothuria forskali*. *International Journal of Biological Macromolecules*, 4, 329–34.

Baldwin, J. G. (1983). Fine structure of the body wall cuticle of females of *Meloidodera charis*,

Atalodera lonicerae, and *Sarisodera hydrophila* (Heteroderidae). *Journal of Nematology*, 15, 370–81.

Baldwin, J. G. (1986). Testing hypotheses of phylogeny of Heteroderidae. In *Cyst Nematodes*, ed. F. Lamberti & C. E. Taylor, pp. 75–100. London: Plenum Press.

Banerjee, S. (1988a). Organization of wing cuticle in *Locusta migratoria* Linnaeus, *Tropidacris cristata* Linnaeus and *Romalea microptera* Beauvais (Orthoptera: Acrididae). *International Journal of Insect Morphology and Embryology*, 17, 313–26.

Banerjee, S. (1988b). The functional significance of wing architecture in acridid Orthoptera. *Journal of Zoology (London)*, 215, 249–67.

Bansal, M. K., Ross, A. S. A., & Bard, J. B. L. (1989). Does chondroitin sulphate have a role to play in the morphogenesis of the chick primary corneal stroma? *Developmental Biology*, 133, 185–95.

Bard, J. (1990). *Morphogenesis: The Cellular and Molecular Processes of Developmental Anatomy*. Cambridge University Press.

Bard, J. B. L., & Higginson, K. (1977). Fibroblast–collagen interactions in the formation of the secondary stroma of the chick cornea. *Journal of Cell Biology*, 74, 816–29.

Barnabus, A. D., Butler, V., & Steinke, T. D. (1977). *Zostera capensis* Setchell. I. Observations on the fine structure of the leaf epidermis. *Zeitschrift für Pflanzenphysiologie*, 85, 417–27.

Barth, F. G. (1969). Die Feinstruktur des Spinnenintegumentes. I. Die Cuticula des Laufbeins adulter häutungsferner Tiere (*Cupiennius salei* Keys). *Zeitschrift für Zellforschung und Mikroskopische Anatomie*, 97, 137–59.

Barth, F. G. (1970). Die Feinstruktur des Spinnenintegumentes. II. Die räumliche Anordnung der Mikrofasern in der lamellierten Cuticula und ihre Beziehung zur Gestalt der Porenkanäle (Cupiennius salei Keys, adult, häutungsfern, Tarsus). *Zeitschrift für Zellforschung und Mikroskopische Anatomie*, 104, 87–106.

Barth, F. G. (1973). Microfiber reinforcement of an arthropod cuticle: laminated composite material in biology. *Zeitschrift für Zellforschung und mikroskopische Anatomie*, 144, 409–33.

Bassot, J. M. (1966). Une forme microtubulaire et paracristalline de réticulum endoplasmique dans les photocytes des Annélides Polynoinae. *Journal of Cell Biology*, 31, 135–45.

Bate, R. H., & East, B. A. (1972). The structure of the ostracod cuticle. *Lethaia*, 5, 177–94.

Bauer, W. D., Talmadge, K. W., Keegstra, K., & Albersheim, P. (1973). The structure of plant cell walls. II. The hemicellulose of the walls of suspension-cultured sycamore cells. *Plant Physiology (Bethesda)*, 51, 174–87.

Bay, E. C. (1965). Instant fish! A new tool for mosquito control? *Pest Control Magazine*, 33, 14–16.

Baynes, S. M. (1972). Light and electron microscope studies on the germination of *Chara* oospores. B.Sc. thesis, University of Bristol, U.K.

Benedict, C. R., & Scott, J. R. (1976). Photosynthetic carbon metabolism of a marine grass. *Plant Physiology (Bethesda)*, 57, 876–80.

Bennet-Clark, H. C. (1976). Energy storage in jumping animals. In *Perspectives in Experimental Biology. Vol. 1: Zoology*, ed. P. Spencer-Davies, pp. 467–79. Oxford: Pergamon Press.

Bhadani, S. N., & Gray, D. G. (1984). Crosslinked cholesteric network from the acrylic acid ester of (hydroxypropyl) cellulose. *Molecular Crystals and Liquid Crystals*, 102, 255–60.

Birch, W. R. (1974). The unusual epidermis of the marine angiosperm *Halophila*. *Flora (Jena)*, 163, 410–14.

Bird, A. F., & Deutsch, K. (1957). The structure of the cuticle of *Ascaris lumbricoides*. *Parasitology*, 47, 319–28.

Blackwell, J., & Weih, M. A. (1980). The structure of chitin-protein complexes: the ovipositor of the ichneumon fly *Megarhyssa*. *Journal of Molecular Biology*, 137, 49–60.

Bogduk, N., & Twomey, L. T. (1987). *Clinical Anatomy of the Lumbar Spine*. London: Churchill Livingstone.

Boissière, J. C. (1987). Ultrastructural relationship between the composition and the structure of the cell wall of the mycobiont of two lichens. In *Progress and Problems in Lichenology in the Eighties*, ed. E. Peveling. Berlin: J. Cramer. *Bibliotheca Lichenologica*, 25, 117–32.

Bonfante-Fasolo, P. (1983). Electron microscopic cytochemical study of cell wall in *Glomus epigaeum* spore. In *Third International Mycological Congress Abstracts*, p. 392. Tokyo.

Bonfante-Fasolo, P., & Grippiolo, R. (1984). Cytochemical and biochemical observations on the cell wall of the spore of *Glomus epigaeum*. *Protoplasma*, 123, 140–51.

Bonfante-Fasolo, P., & Vian, B. (1984). Wall texture in the spore of a vesicular-arbuscular mycorrhizal fungus. *Protoplasma*, 120, 51–60.

Bonfante-Fasolo, P., Vian, B., & Testa, B. (1986). Ultrastructural localization of chitin in the cell wall of a fungal spore. *Biology of the Cell*, 57, 265–70.

Bordereau, C. (1982). Ultrastructure and

formation of the physogastric termite queen cuticle. *Tissue and Cell*, 14, 371–96.

Bouligand, Y. (1965). Sur une architecture torsadée répandue dans de nombreuses cuticules d'arthropodes. *Comptes Rendus Hebdomadaires des Séances de l'Académie des Sciences*, 261, 3665–8.

Bouligand, Y. (1969). Sur l'existence de 'pseudomorphoses cholestériques' chez divers organismes vivants. *Le Journal de Physique*, 30 (Suppl. C4), 90–103.

Bouligand, Y. (1971). Les orientations fibrillaires dans le squelette des arthropodes. *Journal de Microscopie* (Paris), 11, 441–72.

Bouligand, Y. (1972). Twisted fibrous arrangements in biological materials and cholesteric mesophases. *Tissue and Cell*, 4, 189–217.

Bouligand, Y. (1978a). Cholesteric order in biopolymers. In *Mesomorphic Order in Polymers and Polymerization in Liquid Crystalline Media*, ed. A. Blumstein, pp. 237–47. Washington: American Chemical Society.

Bouligand, Y. (1978b). Liquid crystals and their analogs in biological systems. *Solid State Physics Supplement*, 14, 259–94.

Bouligand, Y. (1981a). Geometry and topology of defects in liquid crystals. In *Physics of Defects*, ed. R. Balian, pp. 667–771. Amsterdam: North Holland.

Bouligand, Y. (1981b). Defects in ordered biological materials. In *Physics of Defects*, ed. R. Balian, pp. 779–811. Amsterdam: North Holland.

Bouligand, Y. (1986). Theory of microtomy artifacts in arthropod cuticle. *Tissue and Cell*, 18, 621–43.

Bouligand, Y., Denèfle, J. P., Lechaire, J. P., & Maillard, M. (1985). Twisted architecture in cell-free assembled collagen gels: study of collagen substrates used for cultures. *Biology of the Cell*, 54, 143–62.

Bouligand, Y., & Giraud-Guille, M. M. (1985). Spatial organization of collagen fibrils in skeletal tissues: analogues with liquid crystals. In *Biology of Invertebrate and Lower Vertebrate Collagens*, ed. A. Bairati & R. Garrone, pp. 115–34. New York: Plenum Press.

Bouligand, Y., & Kléman, M. (1970). Paires de disinclinaisons hélicoidales dans les cholestériques. *Le Journal de Physique*, 31, 1041–54.

Bouligand, Y., & Livolant, F. (1984). The organization of cholesteric spherulites. *Le Journal de Physique*, 45, 1899–923.

Boyd, J. D. (1985). *Biophysical Control of Microfibril Orientation in Plant Cell Walls. Aquatic and Terrestrial Plants Including Trees*. Dordrecht: Nijhoff & Junk.

Brodie, A. E. (1970). Development of the cuticle in the rotifer *Asplanchna brightwelli*. *Zeitschrift für Zellforschung und Mikroskopische Anatomie*, 105, 515–25.

Browaeys-Poly, E. (1991). Extracellular matrix of the regeneration chamber and plasma membranes of the epidermis during leg regeneration in an insect *Carausius morosus*. *Tissue and Cell*, 23, 41–55.

Brown, S. C., & McGee-Russell, S. (1971). *Chaetopterus* tubes: ultrastructural architecture. *Tissue and Cell*, 3, 65–70.

Brown, T. (1987). An ultrastructural study of the ecdysial fracture region in *Romalea microptera*. B.Sc. thesis, University of Bristol, U.K.

Brummell, D. A., & Hall, J. L. (1985). The role of cell wall synthesis in sustained auxin-induced growth. *Physiologia Plantarum*, 63, 406–12.

Buchala, A. J., & Meier, H. (1972). Hemicelluloses from the stalk of *Cyperus papyrus*. *Phytochemistry*, 11, 3275–8.

Burke, J. M., & Ross, R. (1975). A radioautographic study of collagen synthesis by earthworm epidermis. *Tissue and Cell*, 7, 631–50.

Callow, M. E., Coughlan, S. J., & Evans, L. U. (1978). The role of Golgi bodies in polysaccharide sulphation in *Fucus* zygotes. *Journal of Cell Science*, 32, 337–56.

Carlstrøm, D. (1957). The crystal structure of α-chitin (poly-N-acetyl-D-glucosamine). *Journal of Biophysical and Biochemical Cytology*, 3, 669–83.

Carpita, N. C. (1985). Tensile strength of cell walls of living cells. *Plant Physiology* (Bethesda), 79, 485–8.

Caveney, S. (1969). Muscle attachment related to cuticle architecture in Apterygota. *Journal of Cell Science*, 4, 541–59.

Caveney, S. (1970). Juvenile hormone and wound modelling of *Tenebrio* cuticle architecture. *Journal of Insect Physiology*, 16, 1087–107.

Caveney, S. (1971). Cuticle reflectivity and optical activity in scarab beetles: the rôle of uric acid. *Proceedings of the Royal Society of London (B), Biological Sciences*, 178, 205–25.

Caveney, S. (1973). Stability of polarity in the epidermis of a beetle, *Tenebrio molitor* L. *Developmental Biology*, 30, 321–35.

Caveney, S., & McIntyre, P. (1981). Design of graded-index lenses in the superposition eyes of scarab beetles. *Philosophical Transactions of the Royal Society of London (B)*, 294, 589–635.

Chafe, S. C. (1970). The fine structure of the collenchyma cell wall. *Planta* (Berlin), 90, 12–21.

Chafe, S. C. (1974). Cell wall structure in the

xylem parenchyma of *Cryptomeria*. *Protoplasma*, 81, 63–76.

Chafe, S. C., & Chauret, G. (1974). Cell wall structure in the xylem parenchyma of trembling aspen. *Protoplasma*, 80, 129–47.

Chafe, S.C., & Doohan, M. E. (1972). Observations on the ultrastructure of the thickened sieve cell wall in *Pinus strobus* L. *Protoplasma*, 75, 67–78.

Chafe, S.C., & Wardrop, A. B. (1972). Fine structural observations on the epidermis. I. The epidermal cell wall. *Planta (Berlin)*, 107, 269–78.

Chambers, T. J., Revell, P. A., Fuller, K., & Athanassou, N. A. (1984). Resorption of bone by isolated rabbit osteoclasts. *Journal of Cell Science*, 66, 383–99.

Chapman, D. M. (1968). Structure, histochemistry and formation of the podocyst and cuticle of *Aurelia aurita*. *Journal of the Marine Biological Association (U.K.)*, 48, 187–208.

Chauvin, G., Rahn, R., & Barbier, R. (1974). Comparaison des oeufs des Lepidopteres *Phalera bucephala* L. (Ceruridae), *Acrolepia assectella* Z. et *Plutella maculipennis* Curt. (Plutellidae): morphologie et ultrastructures particulieres du chorion au contact du support vegetal. *International Journal of Insect Morphology and Embryology*, 3, 247–56.

Chevone, B. I., & Richards, A. G. (1977). Ultrastructural changes in intersegmental cuticle during rotation of the terminal abdominal segments in a mosquito. *Tissue and Cell*, 9, 241–54.

Chihara, C. J., Silvert, D. J., & Fristrom, J. W. (1982). The cuticle proteins of *Drosophila melanogaster*: stage specificity. *Developmental Biology*, 89, 379–88.

Chothia, C., Levitt, M., & Richardson, D. (1977). Structure of proteins. Packing of α-helices and pleated sheets. *Proceedings of the National Academy of Sciences, USA*, 74, 4130–4.

Chou, T. W., McCullough, R. L., & Pipes, R. B. (1986). Composites. *Scientific American*, 255, 166–77.

Chu, H., Norris, D. M., & Carlson, S. D. (1975). Ultrastructure of the compound eye of the diploid female beetle, *Xyleborus ferrugineus*. *Cell and Tissue Research*, 165, 23–6.

Clark, R. B., & Cowey, J. B. (1958). Factors controlling the change of shape of certain nemertean and turbellarian worms. *Journal of Experimental Biology*, 35, 731–48.

Clément, P. (1969). Premières observations sur l'ultrastructure comparée des téguments de rotifères. *Vie et Milieu*, 20, 461–82.

Clément, P. (1977). Ultrastructural research in rotifers. *Archiv für Hydrobiologie*, 8, 270–97.

Cliff, G. M., & Baldwin, J. G. (1985). Fine structure of the body wall cuticle of females of eight genera of Heteroderidae. *Journal of Nematology*, 17, 286–96.

Compère, P., & Goffinet, G. (1987). Ultrastructural shape and three-dimensional organization of the intracuticular canal systems in the mineralized cuticle of the green crab *Carcinus maenas*. *Tissue and Cell*, 19, 839–57.

Coulombre, A. J. (1965). Problems in corneal morphogenesis. *Advances in Morphogenesis*, 4, 81–109.

Coulombre, A. J., & Coulombre, J. L. (1961). The development of the structural and optical properties of the cornea. In *The Structure of the Eye*, ed. G. K. Smelser, pp. 405–19. New York: Academic Press.

Coulombre, J., & Coulombre, A. (1975). Corneal development. V. Treatment of five-day-old embryos of domestic fowl with 6-diazo-5-oxo-L-norleucine (DON). *Developmental Biology*, 45, 291–303.

Cowey, J. B. (1952). The structure and function of the basement membrane–muscle system in *Amphiporus lactiflorens* (Nemertea). *Quarterly Journal of Microscopical Science*, 93, 1–15.

Cox, D. L., & Willis, J. H. (1985). The cuticular proteins of *Hyalophora cecropia* from different anatomical regions and metamorphic stages. *Insect Biochemistry*, 15, 349–62.

Cox, G., & Juniper, B. (1973). Electron microscopy of cellulose in entire tissue. *Journal of Microscopy (Oxford)*, 97, 343–55.

Credland, P. F. (1983). Organization of the cuticle of an aquatic fly larva. *Tissue and Cell*, 15, 477–88.

Crick, F. H. C. (1953). The packing of α-helices in simple coiled-coils. *Acta Crystallographica*, 6, 689–97.

Currey, J. D. (1962). Strength of bone. *Nature (London)*, 195, 513–14.

Dalingwater, J. E. (1975). The reality of arthropod cuticular laminae. *Cell and Tissue Research*, 163, 411–13.

Darke, W. (1986). Asymmetry and collagen fibre orientation in the basal plate of flatfish (Pleuronectidae) scales. B.Sc. thesis, University of Bristol, U.K.

Darvill, A., McNeil, M., Albersheim, P., & Delmer, D. P. (1980). The primary cell walls of flowering plants. In *The Biochemistry of Plants*, Vol. 1, ed. N. E. Tolbert, pp. 91–162. New York: Academic Press.

David, W. A. L. (1967). The physiology of the insect integument in relation to the invasion of pathogens. In *Insects and Physiology*, ed. J. W. L. Beament & J. E. Treherne, pp. 17–35. Edinburgh: Oliver & Boyd.

Delachambre, J. (1975). Les variations de l'architecture dans la cuticule abdominale chez *Tenebrio molitor* L. (Insecta, Coleoptera). *Tissue and Cell*, 7, 669–76.

Dennell, R. (1973). The structure of the cuticle of the shore crab *Carcinus maenas* (L.). *Zoological Journal of the Linnean Society*, 52, 159–63.

Dennell, R. (1974). The cuticle of the crabs *Cancer pagurus* L. and *Carcinus maenas* (L.). *Zoological Journal of the Linnean Society*, 54, 241–5.

Dennell, R. (1976). The fine structure of the cuticle of some Phasmida. In *The Insect Integument*, ed. H. R. Hepburn, pp. 177–92. Amsterdam: Elsevier.

Dentini, M., Crescenzi, V., & Blasi, D. (1984). Conformational properties of xanthan derivatives in dilute aqueous solution. *International Journal of Biological Macromolecules*, 6, 93–8.

Derksen, J. (1986). Cytoskeletal control of cellulose microfibril deposition. In *Cell Walls '86: Proceedings of the Fourth Cell Wall Meeting, Paris*, ed. B. Vian, D. Reis, & R. Goldberg, pp. 34–7. Paris: Université Pierre et Marie Curie.

Deshpande, B. P. (1976). Observations on the fine structure of plant cell walls. II. The microfibrillar framework of the parenchymatous cell wall in *Cucurbita*. *Annals of Botany (London)*, 40, 439–42.

De Vos, L. (1972). Fibres géantes de collagène chez l'éponge *Ephydatia fluviatilis*. *Journal de Microscopie*, 15, 247–57.

Diamant, J., Keller, A., Baer, E., Litt, M., & Arridge, R. G. C. (1972). Collagen: ultrastructure and its relation to mechanical properties as a function of ageing. *Proceedings of the Royal Society of London (B)*, 180, 293–315.

Dickson, M. R., & Mercer, E. H. (1967). Fine structural changes accompanying desiccation in *Philodina roseola* (Rotifera). *Journal de Microscopie*, 6, 331–48.

Digby, P. S. B. (1968). The mechanism of calcification in the molluscan shell. In *Studies in the Structure, Physiology and Ecology of Molluscs. Symposia of the Zoological Society of London*, 22, ed. V. Fretter, pp. 93–107. London: Academic Press.

Dlugosz, J., Gathercole, L. J., & Keller, A. (1979). Cholesteric analogue packing of collagen fibrils in the Cuvierian tubules of *Holothuria forskali* (Holothuroidea, Echinodermata). *Micron*, 10, 81–7.

Doohan, M. E., & Newcomb, E. H. (1976). Leaf ultrastructure and ^{13}C values of three seagrasses from the Great Barrier Reef. *Australian Journal of Plant Physiology*, 3, 9–23.

Duke-Elder, S., & Wybar, K. C. (1961). *The Anatomy of the Visual System. System of Ophthalmology, Vol. 2*. London: Henry Kimpton.

Dunning, C. E. (1968). Cell wall morphology of long leaf pine latewood. *Wood Science*, 1, 65–76.

Dutkowski, A. B., Oberlander, H., & Leach, C. E. (1977). Ultrastructure of cuticle deposited in *Plodia interpunctella* wing discs after various β-ecdysone treatments *in vitro*. *Wilhelm Roux's Archives*, 183, 155–64.

Emons, A. M. C. (1982). Microtubules do not control microfibril orientation in a helicoidal cell wall. *Protoplasma*, 113, 85–7.

Emons, A. M. C. (1986). Cell wall texture in root hairs of the genus *Equisetum*. *Canadian Journal of Botany*, 64, 2201–6.

Emons, A. M. C., & van Maaren, N. (1987). Helicoidal cell wall texture in root hairs. *Planta (Berlin)*, 170, 145–51.

Erhardt, H. V. (1978). Electronenmikroskopische Untersuchungen an den Eihüllen von *Lutjanus analis* (Cuvier and Valenciennes 1828) (Lutjanidae, Perciformes, Pisces). *Biologisches Zentralblatt*, 97, 181–7.

Erickson, R. O. (1982). Mathematical models of plant morphogenesis. *Acta Biotheoretica*, 31A, 132–51.

Fergason, J. L. (1968). Liquid crystals in nondestructive testing. *Applied Optics*, 7, 1729–37.

Filshie, B. K. (1970). The fine structure and deposition of the larval cuticle of the sheep blowfly (*Lucilia cuprina*). *Tissue and Cell*, 3, 479–98.

Filshie, B. K., & Campbell, I. C. (1984). Design of an insect cuticle associated with osmoregulation: the porous plates of chloride cells in a mayfly nymph. *Tissue and Cell*, 16, 789–803.

Filshie, B. K., & Hadley, N. F. (1979). Fine structure of the cuticle of the desert scorpion *Hadrurus arizonensis*. *Tissue and Cell*, 11, 249–62.

Filshie, B. K., & Smith, D. S. (1980). A proposed solution to a fine-structural puzzle: the organization of gill cuticle in a crayfish (*Panulirus*). *Tissue and Cell*, 12, 209–26.

Fitton-Jackson, S. (1968). The morphogenesis of collagen. In *Treatise on Collagen, Vol. 2B*, ed. B. S. Gould, pp. 1–66. London: Academic Press.

Flory, P. J. (1961). Phase changes in proteins and polysaccharides. *Journal of Polymer Science*, 69, 105–28.

Fraenkel, G., & Rudall, K. M. (1947). The

structure of insect cuticles. *Proceedings of the Royal Society of London (B)*, 134, 111–43.

Frank, F. C. (1958). On the theory of liquid crystals. *Discussions of the Faraday Society*, 25, 19–28.

Frey-Wyssling, A. (1976). *The Plant Cell Wall*. Encyclopedia of Plant Anatomy, III, 4, pp. 1–294. Berlin: Borntraeger.

Fry, S. C. (1988). *The Growing Plant Cell Wall. Chemical and Metabolic Analysis*, pp. 1–352. Harlow, U.K.: Longman.

Furneaux, P. J. S., James, C. R., & Potter, S. A. (1969). The egg shell of the house cricket (*Acheta domesticus*): an electron microscope study. *Journal of Cell Science*, 5, 227–49.

Furneaux, P. J. S., & Mackay, A. L. (1972). Crystalline protein in the chorion of insect egg shells. *Journal of Ultrastructural Research*, 38, 343–59.

Furneaux, P. J. S., & Mackay, A. L. (1976). The composition, structure and formation of the chorion and the vitelline membrane of the insect egg shell. In *The Insect Integument*, ed. H. R. Hepburn, pp. 157–76. New York: Elsevier.

Gaill, F. (1990). Geometry of biological interfaces: the collagen networks. *Journal de Physique*, 23, 169–82.

Gaill, F., & Bouligand, Y. (1985). Long pitch helices in invertebrate collagens. In *Biology of Invertebrate and Lower Vertebrate Collagens*, ed. A. Bairati & R. Garrone, pp. 267–74. New York: Plenum Press.

Gaill, F., & Bouligand, Y. (1987). Alternating positive and negative twist of polymers in an invertebrate integument. *Molecular Crystals and Liquid Crystals*, 153, 31–41.

Gaill, F., Herbage, D., & Lepescheux, L. (1988). Cuticle structure and competition of two invertebrates of hydrothermal vents: *Alvinella pompejana* and *Riftia pachyptila*. *Oceanologica Acta (European Journal of Oceanology)*, Special Vol. 8, 155–9.

Gaill, F., & Hunt, S. (1986). Tubes of deep sea hydrothermal vent worms *Riftia pachyptila* (Vestimentifera) and *Alvinella pompejana* (Annelida). *Marine Ecology – Progress Series*, 34, 267–74.

Gardner, K. H., & Blackwell, J. (1974). The structure of native cellulose. *Biopolymers*, 13, 1975–81.

Garrone, R., Vacelet, J., Pavans, M., de Ceccatty, P., Junqua, S., Robert, L., & Huc, A. (1973). Une formation collagène particulière: les filaments des Eponges Cornées Ircinia. Etude ultrastructurale, physicochimique. *Journal de Microscopie*, 17, 241–51.

Gathercole, L. J., Barnard, K., & Atkins, E. D. T. (1989). Molecular organization of type IV collagen: polymer liquid crystal-like aspects. *International Journal of Biological Macromolecules*, 11, 335–88.

Gebhardt, W. (1906). Ueber funktionell wichtige Anordnungs weisen der grösseren und feineren Bauelemente der Wilbertierknochens. *Roux Archives Entwicklungs Mechanik*, 20, 187–322.

Gharagozlou-van-Ginneken, I. D., & Bouligand, Y. (1973). Ultrastructures tegumentaires chez un crustace copepode *Cletocamptus retrogressus*. *Tissue and Cell*, 5, 413–39.

Giddings, T. H., & Staehelin, L. A. (1988). Spatial relationship between microtubules and plasma-membrane rosettes during the deposition of primary wall microfibrils in *Closterium* sp. *Planta (Berlin)*, 173, 22–30.

Giraud, M. M., Castanet, J., Meunier, F. J., & Bouligand, Y. (1978). The fibrous structure of coelacanth scales: a twisted plywood. *Tissue and Cell*, 10, 671–86.

Giraud-Guille, M. M. (1986). Direct visualization of microtomy artefacts in sections of twisted fibrous extracellular matrices. *Tissue and Cell*, 18, 603–20.

Giraud-Guille, M. M. (1987). Cholesteric twist of collagen *in vivo* and *in vitro*. *Molecular Crystals and Liquid Crystals*, 153, 15–30.

Giraud-Guille, M. M. (1988). Twisted plywood architecture of collagen fibrils in human compact bone osteons. *Calcified Tissue International*, 42, 167–80.

Giraud-Guille, M. M. (1989). Liquid crystalline phases of sonicated type I collagen. *Biology of the Cell*, 67, 97–101.

Gordon, H., & Winfree, A. T. (1978). A single spiral artefact in arthropod cuticle. *Tissue and Cell*, 10, 39–50.

Gordon, J. E. (1988). *The Science of Structures and Materials*. New York: Scientific American Library.

Götting, K. J. (1965). Die Feinstruktur der hüllschichten reifender Oocyten von *Agonus cataphractus* L. (Teleostei, Agonidae). *Zeitschrift für Zellforschung und Mikroskopische Anatomie*, 66, 405–14.

Götting, K. J. (1967). Der Follikel und die peripheren Strukturen der Oocyten der Teleosteer und Amphibien. *Zeistschrift für Zellforschung und Mikroskopische Anatomie*, 79, 481–91.

Gray, D. G. (1983). Liquid crystalline cellulose derivatives. In *Proceedings of the Ninth Cellulose Conference*, ed. A. Sarko. *Journal of Applied Polymer Science (Applied Polymer Symposia)*, 37, 179–92.

Grierson, J. P., & Neville, A. C. (1981). Helicoidal

architecture of fish eggshell. *Tissue and Cell*, 13, 819–30.

Griffith, C. M., & Lai-Fook, J. (1986). Structure and formation of the chorion in the butterfly *Calpodes*. *Tissue and Cell*, 18, 589–601.

Grigonis, G. J., & Soloman, G. B. (1976). *Capillaria hepatica*: fine structure of the egg-shell. *Experimental Parasitology*, 40, 286–97.

Gross, J., & Piez, K. A. (1960). The nature of collagen. I. Invertebrate collagens. In *Calcification in Biological Systems*, ed. R. F. Sognnaes, pp. 395–409. Symposia of the American Association for the Advancement of Science.

Gubb, D. C. (1975). A direct visualization of helicoidal architecture in *Carcinus maenas* and *Halocynthia papillosa* by scanning electron microscopy. *Tissue and Cell*, 7, 19–32.

Gubb, D. C. (1976). Helicoidal architecture and the control of morphogenesis in extracellular matrices. Ph.D. thesis, University of Bristol, U.K.

Gupta, B. L., & Little, C. (1970). Studies on Pogonophora. 4. Fine structure of the cuticle and epidermis. *Tissue and Cell*, 2, 637–96.

Gupta, B. L., & Little, C. (1975). The phylogeny and systematic position of Pogonophora. *Zeitschrift für Zoologische Systematik und Evolutionforschung*, 2, 45–63.

Gurdon, J. B. (1974). *The Control of Gene Expression in Animal Development*. Oxford: Oxford University Press.

Hadley, N. F. (1981). Fine structure of the cuticle of the black widow spider with reference to surface lipids. *Tissue and Cell*, 13, 805–17.

Halcrow, K. (1978). Modified pore canals in the cuticle of *Gammarus* (Crustacea, Amphipoda). A study by scanning and transmission electron microscopy. *Tissue and Cell*, 10, 659–70.

Halcrow, K. (1985). The fine structure of the pore canals of the talitrid amphipod *Hyale nilssoni* Rathke. *Journal of Crustacean Biology*, 5, 606–15.

Halcrow, K. (1988). Absence of epicuticle from the repair cuticle produced by four Malacostracan crustaceans. *Journal of Crustacean Biology*, 8, 346–54.

Halcrow, K., & Smith, J. C. (1986). Wound closure in the crab *Carcinus maenas* (L.). *Canadian Journal of Zoology*, 64, 2770–8.

Hamodrakas, S. J. (1984). Twisted β-pleated sheet: the molecular conformation which possibly dictates the formation of the helicoidal architecture of several proteinaceous eggshells. *International Journal of Biological Macromolecules*, 6, 51–3.

Hamodrakas, S. J., Asher, S. A., Mazur, G. D., Regier, J. C., & Kafatos, F. C. (1982a). Laser Raman studies of protein conformation in the silkmoth chorion. *Biochimica et Biophysica Acta*, 703, 216–22.

Hamodrakas, S. J., Bosshard, H. E., & Carlson, C. N. (1988). Structural models of the evolutionarily conservative central domain of silk-moth chorion proteins. *Protein Engineering*, 2, 201–7.

Hamodrakas, S. J., Jones, C. W., & Kafatos, F. C. (1982b). Secondary structure predictions for silkmoth chorion proteins. *Biochimica et Biophysica Acta*, 700, 42–51.

Hamodrakas, S. J., Kamitsos, E. I., & Papadopoulou, P. G. (1987). Laser-Raman and infrared spectroscopic studies of protein conformation in the eggshell of the fish *Salmo gairdneri*. *Biochimica et Biophysica Acta*, 713, 163–9.

Hamodrakas, S. J., Kamitsos, E. I., & Papanikolaou, A. (1984). Laser-Raman spectroscopic studies of the eggshell (chorion) of *Bombyx mori*. *International Journal of Biological Macromolecules*, 6, 333–6.

Hamodrakas, S. J., Margaritis, L. H., Papassideri, I., & Fowler, A. (1986). Fine structure of the silkmoth *Antheraea polyphemus* chorion as revealed by X-ray diffraction and freeze-fracturing. *International Journal of Biological Macromolecules*, 8, 237–42.

Hamodrakas, S. J., Paulson, J. R., Rodakis, G. C., & Kafatos, F. C. (1983). X-ray diffraction studies of a silkmoth chorion. *International Journal of Biological Macromolecules*, 5, 149–53.

Hanic, L. A., & Craigie, J. S. (1969). Studies on the algal cuticle. *Journal of Phycology*, 5, 89–102.

Harada, H. (1965). Ultrastructure of angiosperm vessels and ray parenchyma. In *Cellular Ultrastructure of Woody Plants*, ed. W. Côté, pp. 235–49. Syracuse, N.Y.: Syracuse University Press.

Harche, M. (1986). Un type original d'architecture pariétale: l'épiderme foliaire de l'Alfa (*Stipa tenacissima*). *Comptes Rendus Hebdomadaires des Séances de l'Académie des Sciences*, 303, 131–4.

Harris, A. K., Stopak, D., & Wild, P. (1981). Fibroblast traction as a mechanism for collagen morphogenesis. *Nature (London)*, 290, 249–51.

Harris, J. E., & Crofton, H. D. (1957). Structure and function in nematodes: internal pressure and cuticular structure in *Ascaris*. *Journal of Experimental Biology*, 34, 116–30.

Haustein, J. (1983). On the ultrastructure of the developing and adult mouse corneal stroma. *Anatomy and Embryology*, 168, 291–305.

Hay, E. D. (1980). Development of the vertebrate cornea. *International Review of Cytology*, 63, 263–322.

Hay, E. D., ed. (1981). *Cell Biology of Extracellular Matrix*. New York: Plenum Press.

Hay, E. D., & Revel, J. P. (1969). *Fine Structure of the Developing Avian Cornea. Monographs in Developmental Biology*, Vol. 1, ed. A. Wolsky & P. Chen. Basel: Karger.

Heath, I. B., & Seagull, R. (1982). Oriented cellulose fibrils and the cytoskeleton: a critical comparison of models. In *The Cytoskeleton in Plant Growth and Development*, ed. C. W. Lloyd, pp. 163–82. London: Academic Press.

Hendricks, G. M., & Hadley, N. F. (1983). Structure of the cuticle of the common house cricket with reference to the location of lipids. *Tissue and Cell*, 15, 761–79.

Hillerton, J. E. (1980). Electron microscopy of fibril–matrix interactions in a natural composite, insect cuticle. *Journal of Materials Science*, 15, 3109–12.

Hinton, H. E. (1981). *Biology of Insect Eggs*, 3 vols. Oxford: Pergamon Press.

Hoffman, L. R., & Hofmann, C. S. (1975). Zoospore formation in *Cylindrocapsa*. *Canadian Journal of Botany*, 53, 439–51.

Hughes, P. M. (1987). Insect cuticular growth layers seen under the scanning electron microscope: a new display method. *Tissue and Cell*, 19, 705–12.

Humphreys, S., & Porter, K. R. (1976). Collagen deposition on a preformed grid. *Journal of Morphology*, 149, 53–71.

Hunt, S. (1971). Comparison of three extracellular structural proteins from the whelk *Buccinum undatum*. The periostracum, operculum and egg capsule. *Comparative Biochemistry and Physiology*, 40B, 37–46.

Hunt, S., & Oates, K. (1970). Fibrous protein ultrastructure of gastropod periostracum (*Buccinum undatum* L.). *Experientia*, 26, 1196–7.

Hunt, S., & Oates, K. (1978). Fine structure and molecular organization of the periostracum in a gastropod mollusc *Buccinum undatum* L. and its relation to similar structural protein systems in other invertebrates. *Philosophical Transactions of the Royal Society of London (B)*, 283, 417–59.

Hunt, S., & Oates, K. (1984). Chitin helicoids accompany protein helicoids in the periostracum of a whelk, *Buccinum*. *Tissue and Cell*, 16, 565–75.

Hunt, S., & Oates, K. (1985). Helicoidal architecture in the periostracum of a terrestrial prosobranch *Pterocyclus latilabrum* Smith. *Journal of Molluscan Studies*, 51, 336–44.

Hynes, R. O., & Destree, A. T. (1978). Relationships between fibronectin (LETS protein) and actin. *Cell*, 15, 875–86.

Itoh, T., & Brown, R. M. (1984). The assembly of cellulose microfibrils in *Valonia macrophysa*. *Planta (Berlin)*, 160, 372–81.

Jackson, M. L. (1985). Insect endocuticular plywoods: their structure and function. B.Sc. thesis, University of Bristol, U.K.

Jagels, R. (1973). Studies of a marine grass, *Thalassia testudinum*. I. Ultrastructure of the osmoregulatory leaf cells. *American Journal of Botany*, 60, 1003–9.

Jarvis, M. C. (1984). Structure and properties of pectin gels in plant cell walls. *Plant Cell and Environment*, 7, 153–64.

Jarvis, M. C. (1992). Self-assembly of plant cell walls. *Plant Cell and Environment*, 15, 1–5.

Jensen, M., & Weis-Fogh, T. (1962). Biology and physics of locust flight. V. Strength and elasticity of locust cuticle. *Philosophical Transactions of the Royal Society of London (B)*, 245, 137–69.

Jeronimidis, G. (1980). Wood, one of nature's challenging composites. In *The Mechanical Properties of Biological Materials*, ed. J. F. V. Vincent & J. D. Currey, pp. 169–82. Symposia of the Society for Experimental Biology, no. 34. Cambridge University Press.

Juniper, B. E., Lawton, J. R., & Harris, P. J. (1981). Cellular organelles and cell-wall formation in fibres from the flowering stem of *Lolium temulentum* L. *New Phytologist*, 89, 609–19.

Kathirithamby, J., Luke, B. M., & Neville, A. C. (1990). The ultrastructure of the preformed ecdysial 'line of weakness' in the puparium cap of *Elenchus tenuicornis* (Kirby) (Insecta: Strepsiptera). *Zoological Journal of the Linnean Society*, 98, 229–36.

Kato, Y., Iki, K., & Matsuda, K. (1981). Cell wall polysaccharides of immature barley plants. II. Characterization of a xyloglucan. *Agricultural and Biological Chemistry*, 45, 2745–53.

Kefalides, N. A., Alper, R., & Clark, C. C. (1979). Biochemistry and metabolism of basement membranes. *International Review of Cytology*, 61, 167–228.

Keller, A., Dlugosz, J., Folkes, M. J., Pedemonte, E., Scalisi, F. P., & Willmouth, F. M. (1971). Macroscopic 'single crystals' of an S-B-S three block copolymer. *Journal de Physique*, 32, C5a, 295–300.

Kenchington, W., & Flower, N. E. (1969). Studies on insect fibrous proteins: the structural protein of the ootheca in the preying mantis, *Sphodromantis centralis* Rehn. *Journal of Microscopy*, 89, 263–81.

Kerr, T., & Bailey, I. W. (1934). The cambium and its derivative tissues. X. Structure, optical properties and chemical composition of the so-called middle lamella. *Journal of the Arnold Arboretum*, 15, 327–49.

Kiely, M. L., & Riddiford, L. M. (1985). Temporal programming of epidermal cell protein synthesis during the larval–pupal transformation on *Manduca sexta*. *Wilhelm Roux's Archives of Developmental Biology*, 194, 325–35.

Knight, D. P., & Hunt, S. (1974a). Molecular and ultrastructural characterization of the egg capsule of the leech *Erpobdella octoculata* L. *Comparative Biochemistry and Physiology (A)*, 47, 871–80.

Knight, D. P., & Hunt, S. (1974b). Fibril structure of collagen in egg capsule of dogfish. *Nature (London)*, 249, 379–80.

Kramer, K. J., Ong, J., & Law, J. H. (1973). Oothecal proteins of the oriental preying mantid, *Tenodera sinensis*. *Insect Biochemistry*, 3, 297–302.

Krenchel, H. (1964). *Fibre Reinforcement*. Copenhagen: Akademisk Forlag.

Kroon, D. B., Veerkamp, T. A., & Loeven, W. A. (1952). X-ray analysis of the process of extension of the wing of the butterfly. *Proceedings of the Koninklijke Nederlandse Akademie van Wetenschappen, Series C, Biological and Medical Sciences*, 55, 209–14.

Lai-Fook, J. (1968). The fine structure of wound repair in an insect (*Rhodnius prolixus*). *Journal of Morphology*, 124, 37–78.

Lamberti, F., & Taylor, C. E. (1986). *Cyst Nematodes*. New York: Plenum Press.

Lang, J. M., Eisinger, W. R., & Green, P. B. (1982). Effects of ethylene on the orientation of microtubules and cellulose microfibrils of pea epicotyl cells with polylamellate cell walls. *Protoplasma*, 110, 5–14.

Lecher, P., & Cachon, M. (1970). La membrane capsulaire et ses différentiations chez les Radiolaires Phaeodariae. In *Proceedings of the 7th International Congress of Electron Microscopy*, 3, 393–4.

Ledbetter, M. C., & Porter, K. R. (1963). A 'microtubule' in plant cell fine structure. *Journal of Cell Biology*, 19, 239–50.

Lee, D. L. (1970). The ultrastructure of the cuticle of adult female *Mermis nigrescens* (Nematoda). *Journal of Zoology, (London)*, 161, 513–18.

Leitch, A. (1986). Studies on living and fossil Charophyte oosporangia. Ph.D. thesis, University of Bristol, U.K.

Leitch, A. (1989). Formation and ultrastructure of a complex, multilayered wall around the oospore of *Chara* and *Lamprothamnium* (Characeae). *British Phycological Journal*, 24, 229–36.

Lemoine, A., Millot, C., Curie, G., & Delachambre, J. (1990). Spatial and temporal variations in cuticle proteins as revealed by monoclonal antibodies. Immunoblotting analysis and ultrastructural immunolocalization in a beetle, *Tenebrio molitor*. *Tissue and Cell*, 22, 177–89.

Lepescheux, L. (1988). Spatial organization of collagen in annelid cuticle: order and defects. *Biology of the Cell*, 62, 17–31.

Levy, S. (1986). Control of microfibril orientation in *Nitella* cell walls. Ph.D. thesis, University of Bristol, U.K.

Levy, S. (1987). A 3-D computer representation of helicoidal superstructures in biological materials. *European Journal of Cell Biology*, 44, 27–33.

Levy, S. (1991). Two separate zones of helicoidally oriented microfibrils are present in the wall of *Nitella* internodes during growth. *Protoplasma*, 163, 145–55.

Levy, S., & Neville, A. C. (1986). Computer methods for 3-dimensional ultrastructural modelling and predictions of dynamic changes in the cell wall during growth. In *Cell Walls '86: Proceedings of the Fourth Cell Wall Meeting, Paris*, ed. B. Vian, D. Reis, & R. Goldberg, pp. 18–19. Paris: Université Pierre et Marie Curie.

Levy, S., York, W. S., Stuike-Prill, R., Meyer, B., & Staehelin, L. A. (1991). Simulations of the static and dynamic molecular conformations of xyloglucan. The role of the fucosylated sidechain in surface-specific sidechain folding. *The Plant Journal*, 1, 195–215.

Liang, C. Y., Bassett, K. H., McGinnes, E. A., & Marchessault, R. H. (1960). Infrared spectra of crystalline polysaccharides. 7. Thin wood sections. *Technical Association of the Pulp and Paper Industry Journal (TAPPI)*, 43, 1017–24.

Liese, W. (1965). The warty layer. In *Cellular Ultrastructure of Woody Plants*, ed. W. A. Côté, pp. 251–69. Syracuse, N.Y.: Syracuse University Press.

Lindberg, B., Mosihuzzaman, M., Nahar, N., Abeysekera, R. M., Brown, R. G., & Willison, J. H. M. (1990). An unusual (4-O-methyl-D-glucurono)-D-xylan isolated from the mucilage of seeds of the quince tree (*Cydonia oblonga*). *Carbohydrate Research*, 207, 307–10.

Livolant, F., Giraud, M. M., & Bouligand, Y. (1978). A goniometric effect observed in sections of twisted fibrous materials. *Biologie Cellulaire*, 31, 159–68.

Lloyd, C. W., ed. (1982). *The Cytoskeleton in Plant Growth and Development*. London: Academic Press.

Lloyd, C. W. (1983). Toward a dynamic helical model for the influence of microtubules on wall patterns in plants. *International Review of Cytology*, 86, 1–51.

Locke, M. (1960). Cuticle and wax secretion in *Calpodes ethlius* (Lepidoptera, Hesperiidae). *Quarterly Journal of Microscopical Science*, 101, 333–8.

Locke, M. (1961). Pore canals and related structures in insect cuticle. *Journal of Biophysical and Biochemical Cytology*, 10, 589–618.

Loder, P. M. J. (1989). An investigation of the structure and growth rates of the cuticle of the phasmid *Heteropteryx dilatata* related to the functions of the different parts of the exoskeleton. B.Sc. thesis, University of Bristol, U.K.

Lönning, S. (1972). Comparative electron microscope studies of teleostean eggs with special reference to the chorion. *Sarsia*, 49, 41–8.

Lukat, R. (1978). Circadian growth layers in the cuticle of behaviourally arhythmic cockroaches (*Blaberus fuscus*, Insecta, Blattoidea). *Experientia*, 24, 477.

Mabb, L. P. (1989). Uptake and effects of herbicides on the lawn moss *Rhytidiadelphus squarrosus*. Ph.D. thesis, University of London.

MacLachlan, G. (1985). Are lipid-linked glycosides required for plant polysaccharide biosynthesis? In *Biochemistry of Plant Cell Walls*, ed. C. T. Brett & J. R. Hillman, pp. 199–220. Society for Experimental Biology, Seminar no. 28. Cambridge University Press.

Mandoli, D. F., & Briggs, W. R. (1984). Fiber optics in plants. *Scientific American*, 251(2), 80–8.

Marchessault, R. H., & Liang, C. Y. (1962). The infrared spectra of crystalline polysaccharides. 8. Xylans. *Journal of Polymer Science*, 59, 357–78.

Maret, G., Milas, M., & Rinaudo, M. (1981). Cholesteric order in aqueous solutions of the polysaccharide xanthan. *Polymer Bulletin*, 4, 291–7.

Mark, R. E. (1967). *Cell Wall Mechanics of Tracheids*. New Haven: Yale University Press.

Matsuzaki, M. (1968). Electron microscopic observations on chorion formation of the silkworm *Bombyx mori*. *Journal of Sericultural Science (Tokyo)*, 37, 483–90.

Mauguin, M. C. (1911). Sur les cristaux liquides de Lehmann. *Bulletin de la Societe Francaise Minéralogie et Crystallographie*, 34, 71–117.

Maurice, D. M. (1957). The structure and transparency of the cornea. *Journal of Physiology (London)*, 136, 263–86.

Mazur, G. D., Regier, J. C., & Kafatos, C. (1982). Order and defects in the silkmoth chorion, a biological analogue of a cholesteric liquid crystal. In *Insect Ultrastructure, Vol. 1*, ed. R. C. King & H. Akai, pp. 150–85. New York: Plenum Press.

Meiboom, S., Sethna, J. P., Anderson, P. W., & Brinkman, W. F. (1981). Theory of the blue phase of cholesteric liquid crystals. *Physics Review Letters*, 46, 1216–19.

Meier, H. (1961). The distribution of polysaccharides in wood fibres. *Journal of Polymer Science*, 51, 11–18.

Meunier, F. J. (1981). 'Twisted plywood' structure and mineralization in the scales of a primitive living fish – *Amia calva*. *Tissue and Cell*, 13, 165–71.

Meyer, J., & Michler, P. (1976). Observations ultrastructurales sur la cellule nourricière du *Synchytrium mercurialis* Fuck, et sur le développement de la paroi du sporocyste. *Marcellia*, 39, 155–68.

Minke, R., & Blackwell, J. (1978). The structure of α-chitin. *Journal of Molecular Biology*, 120, 167–81.

Mizuta, S. (1985). Assembly of cellulose synthesizing complexes on the plasma membrane of *Boodlea coacta*. *Plant and Cell Physiology*, 26, 1443–53.

Mizuta, S., Kurogi, U., Okuda, K., & Brown, R. M. (1989). Microfibrillar structure, cortical microtubule arrangement and the effect of Amiprophosmethyl on microfibril orientation in the thallus cells of the filamentous green alga, *Chamaedoris orientalis*. *Annals of Botany*, 64, 383–94.

Mizuta, S., & Wada, S. (1981). Microfibrillar structure of growing cell wall in a coenocytic green alga *Boergesenia forbesii*. *Botanical Magazine (Tokyo)*, 94, 343–53.

Mizuta, S., & Wada, S. (1982). Effects of light and inhibitors on polylamellation and shift of microfibril orientation in *Boergesenia* cell wall. *Plant and Cell Physiology*, 23, 257–64.

Mora, F., Ruel, K., Comtat, J., & Joseleau, J. P. (1986). Aspects of native and redeposited xylans at the surface of cellulose microfibrils. *Holzforschung*, 40, 85–91.

Morris, J. E., & Afzelius, B. A. (1967). The structure of the cell and outer membranes in encysted *Artemia salina* embryos during cryptobiosis and development. *Journal of Ultrastructural Research*, 20, 244–59.

Morritz, K., & Storch, V. (1970). Ueber den Aufbau des Integumentes der Priapuliden und der Sipunculiden (*Priapulus caudatus* Lamarck, *Phascolion strombi*, Montagu). *Zeitschrift für Zellforschung und Mikroskopische Anatomie*, 105, 55–64.

Mosiniak, M. (1986). Texture of cell walls in

Papyrus and behaviour in writing material. In *Cell Walls '86: Proceedings of the Fourth Cell Wall Meeting, Paris,* ed. B. Vian, D. Reis, & R. Goldberg, pp. 74–5. Paris: Université Pierre et Marie Curie.

Mosiniak, M., & Roland, J.-C. (1986). Variations spontanées du rhythme d'assemblage des parois cellulaires à texture hélicoidale: l'exemple du papyrus. *Annales des Sciences Naturelles Botanique,* 7, 175–212.

Mosiniak-Bessoles, M. (1987). Contribution a l'étude des oscillations spontanées de l'assemblage cellulosique dans les parois des cellules vegetales. Doctoral thesis, Université Pierre et Marie Curie, Paris.

Mosse, B. (1970). Honey-colored, sessile *Endogone* spores. III. Wall structure. *Archiv für Mikrobiologie,* 74, 146–59.

Mueller, S. C., & Brown, R. M. (1980). Evidence for an intramembrane component associated with a cellulose microfibril-synthesizing complex in higher plants. *Journal of Cell Biology,* 84, 315–26.

Mutvei, H. (1974). SEM studies on arthropod exoskeletons. Part I. Decapod crustaceans, *Homarus gammarus* L. and *Carcinus maenas* (L.). *Bulletin of the Geological Institutions of the University of Uppsala, New Series,* 4, 73–80.

Nadel, M. R., Goldsmith, M. R., Goplerud, J., & Kafatos, F. C. (1980). Specific protein synthesis in cellular differentiation. V. A secretory defect of chorion formation in the Gr^{col} mutant of *Bombyx mori. Developmental Biology,* 75, 41–58.

Nadol, J. B., Gibbins, J. R., & Porter, K. R. (1969). A reinterpretation of the structure and development of the basement lamella: an ordered array of collagen in fish skin. *Developmental Biology,* 20, 304–31.

Nanko, H., Saiki, H., & Harada, H. (1978). Cell wall structure of the sclereids in the secondary phloem of *Populus eurameniana. Mokuzaï Gakkaishi (Journal of the Japanese Wood Research Society),* 24, 362–8.

Neville, A. C. (1963a). Daily growth layers for determining the age of grasshopper populations. *Oikos (Acta Ecologica Scandinavica),* 14, 1–8.

Neville, A. C. (1963b). Growth and deposition of resilin and chitin in locust rubberlike cuticle. *Journal of Insect Physiology,* 9, 265–78.

Neville, A. C. (1965a). Chitin lamellogenesis in locust cuticle. *Quarterly Journal of Microscopical Science,* 106, 269–86.

Neville, A. C. (1965b). Circadian organization of chitin in some insect skeletons. *Quarterly Journal of Microscopical Science,* 106, 315–25.

Neville, A. C. (1967a). Daily growth layers in animals and plants. *Biological Reviews,* 42, 421–41.

Neville, A. C. (1967b). A dermal light sense influencing skeletal structure in locusts. *Journal of Insect Physiology,* 13, 933–9.

Neville, A. C. (1967c). Chitin orientation in cuticle and its control. In *Advances in Insect Physiology,* Vol. 4, ed. J. W. L. Beament, J. E. Treherne, & V. B. Wigglesworth, pp. 213–86. London: Academic Press.

Neville, A. C. (1970). Cuticle ultrastructure in relation to the whole insect. In *Insect Ultrastructure,* ed. A. C. Neville, pp. 17–39. Symposium of the Royal Entomological Society of London, Vol. 5. Oxford: Blackwell.

Neville, A. C. (1975a). *Biology of the Arthropod Cuticle.* Berlin: Springer-Verlag.

Neville, A. C. (1975b). Structural hierarchy of insect cuticle. In *Structure of Fibrous Biopolymers,* ed. E. D. T. Atkins, & A. Keller, pp. 259–70. Colston Papers, no. 26. London: Butterworth.

Neville, A. C. (1976). *Animal Asymmetry.* Studies in Biology, no. 67. London: Arnold.

Neville, A. C. (1977). Metallic gold and silver colours in some insect cuticles. *Journal of Insect Physiology,* 23, 1267–74.

Neville, A. C. (1978). *The Biology of the Arthropod Cuticle.* Carolina Biology Readers, no. 103. Burlington: Carolina Biological Supply Company.

Neville, A. C. (1980). Optical methods in cuticle research. In *Cuticle Techniques in Arthropods,* ed. T. A. Miller, pp. 45–89. Berlin: Springer-Verlag.

Neville, A. C. (1981). Cholesteric proteins. *Molecular Crystals and Liquid Crystals,* 76, 279–86.

Neville, A. C. (1983). Daily cuticular growth layers and the teneral stage in adult insects: a review. *Journal of Insect Physiology,* 29, 211–19.

Neville, A. C. (1984). Cuticle: organization. In *Biology of the Integument. I. Invertebrates,* ed. J. Bereiter-Hahn, A. G. Matoltsy, & K. S. Richards, pp. 611–25. Berlin: Springer-Verlag.

Neville, A. C. (1985a). Molecular and mechanical aspects of helicoid development in plant cell walls. *BioEssays,* 3, 4–8.

Neville, A. C. (1985b). Insect cuticles compared with plant cell walls. *Antenna,* 9, 171–5.

Neville, A. C. (1988a). A pipe-cleaner molecular model for morphogenesis of helicoidal plant cell walls based on hemicellulose complexity. *Journal of Theoretical Biology,* 131, 243–54.

Neville, A. C. (1988b). The need for a constraining layer in the formation of monodomain helicoids in a wide range of biological structures. *Tissue and Cell,* 20, 133–43.

Neville, A. C. (1988c). The helicoidal arrangement of microfibrils in some algal cell walls. In *Progress in Phycological Research*, Vol. 6, ed. F. E. Round & D. J. Chapman, pp. 1–21. Bristol, U.K.: Biopress Ltd.

Neville, A. C., & Berg, C. W. (1971). Cuticle ultrastructure of a Jurassic crustacean (*Eryma stricklandi*). *Palaeontology*, 14, 201–5.

Neville, A. C., & Caveney, S. (1969). Scarabaeid beetle exocuticle as an optical analogue of cholesteric liquid crystals. *Biological Reviews*, 44, 531–62.

Neville, A. C., Gubb, D. C., & Crawford, R. M. (1976). A new model for cellulose architecture in some plant cell walls. *Protoplasma*, 90, 307–17.

Neville, A. C., & Levy, S. (1984). Helicoidal orientation of cellulose microfibrils in *Nitella opaca* internode cells: ultrastructure and computed theoretical effects of strain reorientation during wall growth. *Planta (Berlin)*, 162, 370–84.

Neville, A. C., & Levy, S. (1985). The helicoidal concept in plant cell wall ultrastructure and morphogenesis. In *Biochemistry of Plant Cell Walls*, ed. C. T. Brett & J. R. Hillman, pp. 99–124. Society for Experimental Biology Seminar, no. 28. Cambridge University Press.

Neville, A. C., & Luke, B. M. (1969a). Molecular architecture of adult locust cuticle at the electron microscope level. *Tissue and Cell*, 1, 355–66.

Neville, A. C., & Luke, B. M. (1969b). A two-system model for chitin-protein complexes in insect cuticles. *Tissue and Cell*, 1, 689–707.

Neville, A. C., & Luke, B. M. (1971a). Form optical activity in crustacean cuticle. *Journal of Insect Physiology*, 17, 519–26.

Neville, A. C., & Luke, B. M. (1971b). A biological system producing a self-assembling cholesteric protein liquid crystal. *Journal of Cell Science*, 8, 93–109.

Neville, A. C., Parry, D. A. D., & Woodhead-Galloway, J. (1976). The chitin crystallite in arthropod cuticle. *Journal of Cell Science*, 21, 73–82.

Neville, A. C., Thomas, M. G., & Zelazny, B. (1969). Pore canal shape related to molecular architecture of arthropod cuticle. *Tissue and Cell*, 1, 183–200.

Newman, P. J. (1990). Silicification and encystment in *Actinophrys sol* (Heliozoa, Protista). Ph.D. thesis, University of Bristol, U.K.

Nicholls, S. P., Gathercole, L. J., Keller, A., & Shah, J. S. (1983). Crimping in rat tail tendon collagen: morphology and transverse mechanical anisotropy. *International Journal of Biological Macromolecules*, 5, 283–8.

Noble-Nesbitt, J. (1963). The fully formed intermoult cuticle and associated structures of *Podura aquatica* (Collembola). *Quarterly Journal of Microscopical Science*, 104, 253–70.

Noguchi, T., & Ueda, K. (1985). Cell walls, plasma membranes, and dictyosomes during zygote maturation of *Closterium ehrenbergii*. *Protoplasma*, 128, 64–71.

Noirot, C., & Noirot-Timothée, C. (1971). Ultrastructure du proctodeum chez le Thysanure *Lepismodes inquilinus* Newman (= *Thermobia domestica* Packard). II. Le sac anal. *Journal of Ultrastructural Research*, 37, 335–50.

Northcote, D. H. (1977). The synthesis and assembly of plant cell walls: possible control mechanisms. In *Cell Surface Reviews*, Vol. 4, ed. G. Poste and G. L. Nicholson, pp. 717–39. Amsterdam: North Holland.

Nuñez, J. A. (1963). Central nervous control of the mechanical properties of the cuticle in *Rhodnius prolixus*. *Nature (London)*, 199, 621–2.

Oberlander, H., Lynn, D. E., & Leach, C. E. (1983). Inhibition of cuticle production in imaginal discs of *Plodia interpunctella* (cultured *in vitro*): effects of colcemid and vinblastine. *Journal of Insect Physiology*, 29, 47–53.

O'Donnell, J. L. (1992). An investigation into the helicoidal rotation of collagen fibres in the cornea of different bird species. B.Sc. thesis, University of Bristol, U.K.

Ougherram, A. (1989). Différenciation des parois cellulaires pendant la maturation de la drupe de l'olivier (*Olea europea* L.): sclérification ou autolyse. *Annales des Sciences Naturelles Botanique et Biologie Vegetale (Paris)*, 13, 77–96.

Overton, J. (1976). Scanning microscopy of collagen in basement lamella of normal and regenerating frog tadpoles. *Journal of Morphology*, 150, 805–23.

Overton, J. (1979). Differential response of embryonic cells to culture on tissue matrices. *Tissue and Cell*, 11, 89–98.

Ozanics, V., Rayborn, M., & Sagun, D. (1977). Observations on the morphology of the developing primate cornea: epithelium, its innervation and anterior stroma. *Journal of Morphology*, 153, 263–98.

Page, D. H. (1976). A note on the cell wall structure of softwood tracheids. *Wood Fiber*, 7, 246–8.

Page, D. H., El-Hosseiny, F., Bidmade, M. L., & Binet, R. (1976). Birefringence and chemical composition of wood pulp fibres. *Applied Polymer Symposia*, 28, 923–9.

Papanikolaou, A. M., Margaritis, L. H., &

Hamodrakas, S. J. (1986). Ultrastructural analysis of chorion formation in the silkmoth *Bombyx mori*. *Canadian Journal of Zoology*, 64, 1158–73.

Parameswaran, N. (1975). Zur Wandstruktur von Sklereiden in einigen Baumrinden. *Protoplasma*, 85, 305–14.

Parameswaran, N., & Liese, W. (1975). On the polylamellate structure of parenchyma wall in *Phyllostachys edulis* Riv. *International Association of Wood Anatomists (IAWA) Bulletin*, 4, 57–8.

Parameswaran, N., & Liese, W. (1980). Ultrastructural aspects of bamboo cells. *Cellulose Chemistry and Technology*, 14, 587–609.

Parameswaran, N., & Liese, W. (1981). Occurrence and structure of polylamellate walls in some lignified cells. In *Cell Walls '81: Proceedings of the Second Cell Wall Meeting, Göttingen*, ed. D. G. Robinson, & H. Quader, pp. 171–88. Stuttgart: Wissenschaftliche Verlagsgesellschaft.

Parameswaran, N., & Liese, W. (1982). Ultrastructural localization of wall components in wood cells. *Holz als Roh und Werkstoff*, 40, 145–55.

Parameswaran, N., & Sinner, M. (1979). Topochemical studies on the wall of beech bark sclereids by enzymatic and acidic degradation. *Protoplasma*, 101, 197–215.

Parker, M. L. (1979). Gravity-regulated growth of collenchymatous bundle cap-cells in the leaf sheath base of the grass *Echinochloa colonum*. *Canadian Journal of Botany*, 57, 2399–407.

Patterson, D. J. (1979). On the organization and classification of the protozoan *Actinophrys sol* Ehrenberg. *Mikrobios*, 26, 165–208.

Patterson, D. J., & Thompson, D. W. (1981). Structure and elemental composition of the cyst wall of *Echinosphaerium nucleofilum* Barrett (Heliozoa, Actinophryida). *Journal of Protozoology*, 28, 188–92.

Pearlmütter, N. L., & Lembi, C. A. (1978). Localization of chitin in algal and fungal cell walls by light and electron microscopy. *Journal of Histochemistry and Cytochemistry*, 26, 782–91.

Pearlmütter, N. L., & Lembi, C. A. (1980). Structure and composition of *Pithophora oedogonia* (Chlorophyta) cell walls. *Journal of Phycology*, 16, 602–16.

Pedersen, K. J. (1966). The organization of the connective tissue of *Discocelides langi* (Turbellaria, Polyclada). *Zeitschrift für Zellforschung und Mikroskopische Anatomie*, 71, 94–117.

Pendland, J. (1979). Ultrastructural characteristics of *Hydrilla* leaf tissue. *Tissue and Cell*, 11, 79–88.

Peng, H. B., & Jaffe, L. F. (1976). Cell-wall formation in *Pelvetia* embryos. A freeze-fracture study. *Planta (Berlin)*, 133, 57–71.

Perry, R. N., Wharton, D. A., & Clarke, A. J. (1982). The structure of the egg-shell of *Globodera rostochiensis* (Nematoda, Tylenchida). *International Journal of Parasitology*, 12, 481–5.

Picken, L. E. R. (1962). *The Organization of Cells and Other Organisms*. Oxford: Clarendon Press.

Plumptre, A. J. (1987). A study of the mechanics of moulting and anatomy of the arthropod cuticle. B.Sc. thesis, University of Bristol, U.K.

Pluymaekers, H. J. (1980). Cell wall texture in root hairs of *Limnobium stoloniferum*. *Ultramicroscopy*, 5, 105–6.

Pluymaekers, H. J. (1982). A helicoidal cell wall texture in root hairs of *Limnobium stoloniferum*. *Protoplasma*, 112, 107–16.

Poklewski-Koziell, S. Z. (1992). A study of the twisted plywood orientation of collagen fibres in avian corneas. B.Sc. thesis, University of Bristol, U.K.

Powell, C. V. L., & Halcrow, K. (1985). Formation of the epicuticle in a marine isopod, *Idotea baltica* (Pallas). *Journal of Crustacean Biology*, 5, 439–48.

Preston, R. D. (1952). *The Molecular Architecture of Plant Cell Walls*. London: Chapman & Hall.

Preston, R. D. (1964). Structural plant polysaccharides. *Endeavour*, 23, 153–9.

Preston, R. D. (1974). *The Physical Biology of Plant Cell Walls*. London: Chapman & Hall.

Preston, R. D. (1982). The case for multinet growth in growing walls of plant cells. *Planta (Berlin)*, 155, 356–63.

Preston, R. D. (1988). Cellulose-microfibril-orienting mechanisms in plant cell walls. *Planta (Berlin)*, 174, 67–74.

Preston, R. D., & Goodman, R. N. (1967). Structural aspects of cellulose microfibril biosynthesis. *Journal of the Royal Microscopical Society*, 88, 513–27.

Probine, M. C., & Barber, N. F. (1966). The structure and plastic properties of the cell wall of *Nitella* in relation to extension growth. *Australian Journal of Biological Science*, 19, 439–57.

Probine, M. E., & Preston, R. D., (1961). Cell growth and the structure of mechanical properties of the cell in internodal cells of *Nitella opaca*. *Journal of Experimental Botany*, 12, 261–82.

Rajaei, H. (1980). Ultrastructure des parois épidermiques au cours de la réaction géotropique de l'hypocotyle de Tournesol (*Helianthus annuus* L.). *Comptes Rendus*

Hebdomadaires des Séances de l'Académie des Sciences, 291, 101–3.

Rayle, D. L. (1973). Auxin-induced hydrogen ion secretion in Avena coleoptiles and its implications. Planta (Berlin), 114, 63–73.

Reed, R., & Rudall, K. M. (1948). Electron microscope studies on the structure of earthworm cuticles. Biochimica et Biophysica Acta, 2, 7–18.

Regier, J. C., & Kafatos, F. C. (1985). Molecular aspects of chorion formation. In Embryogenesis and Reproduction: Comprehensive Insect Physiology, Biochemistry and Pharmacology, Vol. 1, ed. G. A., Kerkut & L. I. Gilbert, pp. 113–51. Oxford: Pergamon Press.

Regier, J. C., Mazur, G. D., Kafatos, F. C., & Paul, M. (1982). Morphogenesis of silkmoth chorion: initial framework formation and its relation to synthesis of specific proteins. Developmental Biology, 92, 159–74.

Regier, J. C., & Vlahos, N. S. (1988). Heterochrony and the introduction of novel modes of morphogenesis during the evolution of moth choriogenesis. Journal of Molecular Evolution, 28, 19–31.

Reis, D. (1978). Précisions cytochimiques sur l'assemblage in vitro des hémicelluloses de l'hypocotyle de soja (Phaseolus aureus Roxb.). Annales des Sciences Naturelles, 19, 163–93.

Reis, D. (1981). Cytochimie ultrastructurale des parois en croissance par extractions ménagées. Effets comparés du diméthylsulfoxyde et de la méthylamine sur le démasquage de la texture. Annales des Sciences Naturelles, Botanique, Paris, 13th Series, 121–36.

Reis, D., Roland, J.-C., Mosiniak, M., Darzens, D., & Vian, B. (1992). The sustained and warped helicoidal pattern of a xylan-cellulose composite: the stony endocarp model. Protoplasma, 166, 21–34.

Reis, D., Roland, J.-C., & Vian, B. (1985). Morphogenesis of twisted cell wall: chronology following an osmotic shock. Protoplasma, 126, 36–46.

Reis, D., Vian, B., Chanzy, H., & Roland, J.-C. (1991). Liquid crystal-type assembly of native cellulose-glucuronoxylans extracted from plant cell wall. Biology of the Cell, 73, 173–8.

Reis, D., Vian, B., & Roland, J.-C. (1978). In vitro and in vivo polysaccharide assembly. Ultrastructural and cytochemical study of growing plant cell wall components. In Proceedings of the Ninth International Congress of Electron Microscopy, Toronto, Vol. 2, pp. 434–5.

Reynolds, S. E. (1973). The plasticization of the abdominal cuticle in Rhodnius. Ph.D. thesis, University of Cambridge.

Reynolds, S. E. (1975). The mechanism of plasticization of the abdominal cuticle in Rhodnius. Journal of Experimental Biology, 62, 81–98.

Richards, K. S. (1984). Annelida. Cuticle. In Biology of the Integument. I. Invertebrates, ed. J. Bereiter-Hahn, A. G. Matoltsy, & K. S. Richards, pp. 310–22. Berlin: Springer-Verlag.

Rieder, N. (1972). Ultrastruktur und Polysaccharidanteile der Cuticula von Triops cancriformis Bosc. (Crustacea, Notostraca) während der Haütungsvorbereitung. Zeitschrift für Morphologie und Okologie der Tiere, 73, 361–80.

Rieger, R. M. (1981). Morphology of the Turbellaria at the ultrastructural level. Hydrobiologia, 84, 213–29.

Roberts, I. N., Lloyd, C. W., & Roberts, K. (1985). Ethylene-induced microtubule reorientations: mediation by helical arrays. Planta (Berlin), 164, 439–47.

Robinson, C. (1958). Surface structures in liquid crystals. In Surface Phenomena in Chemistry and Biology, ed. J. R. Danielli, K. G. A. Pankhurst, & A. C. Riddiford, pp. 133–9. London: Pergamon Press.

Robinson, C. (1961). Liquid-crystalline structures in polypeptide solutions. Tetrahedron, 13, 219–34.

Robinson, C. (1966). The cholesteric phase in polypeptide solutions and biological structures. Molecular Crystals, 1, 467–94.

Robinson, D. G., & Herzog, W. (1977). Structure, synthesis and orientation of microfibrils. III. A survey of the action of microtubule inhibitors on microtubules and microfibril orientation in Oocystis solitaria. Cytobiologie, 15, 463–74.

Roelofsen, P. A., & Houwink, A. L. (1953). Architecture and growth of the primary cell wall in some plant hairs and in the Phycomyces sporangiophore. Acta Botanica Neerlandica, 2, 218–25.

Roland, J.-C. (1981). Comparison of arced patterns in growing and non-growing polylamellate cell walls of higher plants. In Cell Walls '81: Proceedings of the Second Cell Wall Meeting, Göttingen, ed. D. G. Robinson & H. Quader, pp. 162–70. Stuttgart: Wissenschaftliche Verlagsgesellschaft.

Roland, J.-C., & Mosiniak, M. (1983). On the twisting pattern, texture and layering of the secondary cell walls of limewood. Proposal of an unifying model. International Association of Wood Anatomists (IAWA) Bulletin, 4, 15–26.

Roland, J.-C., Reis, D., Mosiniak, M., & Vian, B. (1982). Cell wall texture along the growth gradient of the mung bean hypocotyl: ordered assembly and dissipative processes. *Journal of Cell Science*, 56, 303–18.

Roland, J.-C., Reis, D., Vian, B., & Roy, S. (1989). The helicoidal plant cell wall as a performing cellulose-based composite. *Biology of the Cell*, 67, 209–20.

Roland, J.-C., Reis, D., Vian, B., Satiat-Jeunemaître, B., & Mosiniak, M. (1987). Morphogenesis of plant cell walls at the supramolecular level: internal geometry and versatility of helicoidal expression. *Protoplasma*, 140, 75–91.

Roland, J.-C., & Vian, B. (1979). The wall of the growing plant cell: its three-dimensional organization. *International Review of Cytology*, 61, 129–66.

Roland, J.-C., Vian, B., & Reis, D. (1977). Further observations on cell wall morphogenesis and polysaccharide arrangement during plant growth. *Protoplasma*, 91, 125–41.

Rosen, B. W. (1974). Fibre composite materials. *American Society of Metallurgists*, 20, 37–47.

Rosin, S. (1946). Ueber Bau und Wachstum der Grenzlamelle der Epidermis bei Amphibienlarven: Analyse einer orthogonalen Fibrillärstruktur. *Revue Suisse de Zoologie*, 53, 133–201.

Rothschild, M., Schlein, Y., Parker, K., Neville, C., & Sternberg, S. (1973). The flying leap of the flea. *Scientific American*, 229(5), 92–100.

Rothschild, M., Schlein, J., Parker, K., Neville, C., & Sternberg, S. (1975). The jumping mechanism of *Xenopsylla cheopsis*. III. Execution of the jump and activity. *Philosophical Transactions of the Royal Society of London (B)*, 271, 499–515.

Rudall, K. M. (1956). Protein ribbons and sheets. *Lectures on the Scientific Basis of Medicine*, 5, 217–30.

Rudall, K. M. (1963). The chitin/protein complexes of insect cuticles. In *Advances in Insect Physiology, Vol. 1*, ed. J. W. L. Beament, J. E. Treherne, & V. B. Wigglesworth, pp. 257–313. London: Academic Press.

Rudall, K. M. (1965). Skeletal structure in insects. In *Aspects of Insect Biochemistry*, ed. T. W. Goodwin, pp. 83–92. Biochemical Society Symposium no. 25. London: Academic Press.

Rudall, K. M. (1967). Conformation in chitin-protein complexes. In *Conformation of Biopolymers, Vol. 2*, ed. G. N. Ramachandran, pp. 751–65. London: Academic Press.

Rudall, K. M. (1968). Comparative biology and biochemistry of collagen. In *Treatise on Collagen, Vol. 2, Part A*, ed. B. S. Gould, pp. 83–137. London: Academic Press.

Rudall, K. M., & Kenchington, W. (1973). The chitin system. *Biological Reviews*, 49, 597–636.

Samulski, E. T., & Tobolsky, A. V. (1969). The liquid crystal phase of poly-γ-benzyl-L-glutamate in solution and in the solid state. *Molecular Crystals and Liquid Crystals*, 7, 433–42.

Sargent, C. (1978). Differentiation of the crossed-fibrillar outer epidermal wall during extension growth in *Hordeum vulgare* L. *Protoplasma*, 95, 309–20.

Sassen, M. M. A., Pluymaekers, H. J., Meekes, H. T. H. M., & De Jong-Emons, A. M. C. (1981). Cell wall texture in root hairs. In *Cell Walls '81: Proceedings of the Second Cell Wall Meeting, Göttingen*, ed. D. G. Robinson & H. Quader, pp. 189–97. Stuttgart: Wissenschaftliche Verlagsgesellschaft.

Satiat-Jeunemaître, B. (1981). Texture et croissance des parois des deux épidermes du coléoptile de maïs. *Annales des Sciences Naturelles, Botanique et Biologie Vegetale*, 13, 163–76.

Satiat-Jeunemaître, B. (1984). Experimental modifications of the twisting and rhythmic pattern in the cell walls of maize coleoptile. *Biology of the Cell*, 51, 373–80.

Satiat-Jeunemaître, B. (1987). Inhibition of the helicoidal assembly of the cellulose-hemicellulose complex by 2,6-dichlorobenzonitrile (DCB). *Biology of the Cell*, 59, 89–96.

Satiat-Jeunemaître, B. (1989). Microtubules, microfibrilles pariétales et morphogenèse végétale: cas des cellules en extension. *Bulletin Société Botanique de France*, 136, 87–98.

Satiat-Jeunemaître, B., & Darzens, D. (1986). In vivo hemicellulose-cellulose equilibrium modifications: effects on helicoidal cell wall assembly. In *Cell Walls '86: Proceedings of the Fourth Cell Wall Meeting, Paris*, ed. B. Vian, D. Reis, & R. Goldberg, pp. 68–9. Université Pierre et Marie Curie.

Sawhney, V. K., & Srivastava, L. M. (1975). Wall fibrils and microtubules in normal and gibberellic-acid-induced growth of lettuce hypocotyl cells. *Canadian Journal of Botany*, 53, 824–35.

Schmidt, W. J. (1924). *Die Bausteine des Tierkörpers in polarisierten Lichte*. Bonn: F. Cohen.

Schnepf, E., & Deichgräber, G. (1983a). Structure and formation of fibrillar mucilages in seed epidermis cells. I. *Collomia grandiflora* (Polemoniaceae). *Protoplasma*, 114, 210–21.

Schnepf, E., & Deichgräber, G. (1983b). Structure and formation of fibrillar mucilages in seed epidermis cells. II. *Ruellia* (Acanthaceae). *Protoplasma*, 114, 222–34.

Schnepf, E., Stein, U., & Deichgräber, G. (1978). Structure, function, and development of the peristome of the moss, *Rhacopilum tomentosum*, with special reference to the problem of microfibril orientation by microtubules. *Protoplasma*, 97, 221–40.

Schulze, F. E. (1863). Ueber die Structur des Tunicatenmantels und sein Verhalten im polarisirten Lichte. *Zeitschrift für Wissenschaftliche Zoologie*, 12, 175–88.

Sellers, T. (1985). *Plywood and Adhesive Technology*. New York: Dekker.

Shepherd, A. M., Clark, S. A., & Dart, P. J. (1972). Cuticle structure in the genus *Heterodera*. *Nematologia*, 18, 1–17.

Silvestri, F. (1903). Acari, Myriapoda et Scorpiones hucusque in Italia reperta. In *Classis Diplopoda. Vol. 1. Segmenta, Tegumentum, Musculi*, pp. 1–272. Vesuviana Portici.

Simmons, P. H. (1989). An investigation into the endocuticle architecture of the orders Coleoptera, Hemiptera and Odonata using the SEM. B.Sc. thesis, University of Bristol, U.K.

Simpson, B. R. C. (1979). The phenology of annual killifishes. In *Fish Phenology: Anabolic Adaptiveness in Teleosts*, ed. P. J. Miller, pp. 243–61. Symposium of the Zoological Society of London, no. 44. London: Academic Press.

Singer, S. J., & Nicolson, G. L. (1972). The fluid mosaic model of the structure of cell membranes. *Science*, 175, 720–30.

Skelly, P. J., & Howells, A. J. (1988). The cuticle proteins of *Lucilia cuprina*: stage specificity and immunological relatedness. *Insect Biochemistry*, 18, 237–48.

Skucas, G. P. (1967). Structure and composition of the resistant sporangial wall in the fungus *Allomyces*. *American Journal of Botany*, 54, 1152–8.

Slifer, E. H., & Sekhon, S. S. (1963). The fine structure of the membranes which cover the egg of the grasshopper, *Melanoplus differentialis*, with special reference to the hydropyle. *Quarterly Journal of Microscopical Science*, 104, 321–34.

Smith, D. S. (1968). *Insect Cells. Their Structure and Function*, Edinburgh: Oliver & Boyd.

Smith, D. S., Telfer, W. H., & Neville, A. C. (1971). Fine structure of the chorion of a moth, *Hyalophora cecropia*. *Tissue and Cell*, 3, 477–98.

Steinbrecht, R. A. (1985). Fine structure and development of the silver and golden cuticle in butterfly pupae. *Tissue and Cell*, 17, 745–62.

Sterba, G., & Müller, H. (1962). Elektronenmikroskopische Untersuchungen über Bildung und Struktur der Eihüllen bei Knochenfisken. I. Die Hüllen junger Oozyten von *Cynolebias belotti* Steindachner (Cyprinodontidae). *Zoologische Jahrbücher. Abteilung für Anatomie und Ontogonie der Tiere*, 80, 65–80.

Sterman, S., & Marsden, J. G. (1968). Bonding organic polymers to glass by silane coupling agents. In *Fundamental Aspects of Fiber Reinforced Plastic Composites*, ed. R. T. Schwartz & H. S. Schwarz, pp. 245–73. New York: Interscience.

Steven, D. M. (1963). The dermal light sense. *Biological Reviews*, 38, 204–40.

Swanson, C. J. (1974). Application of thin shell theory to helically-wound fibrous cuticles. *Journal of Theoretical Biology*, 43, 293–304.

Tachibana, T., & Kambara, H. (1967). Enantiomorphism in the super helices of poly-γ-benzyl-glutamate. *Kolloid-Zeitschrift und Zeitschrift für Polymere*, 219, 40–2.

Takabe, K., Fujita, M., Harada, H., & Saiki, H. (1984). Incorporation of the label from ^{14}C glucose into cell-wall components during the maturation of *Cryptomeria* tracheids. *Mokuzaï Gakkaishi (Journal of the Japan Wood Research Society)*, 30, 103–9.

Takeda, K., & Shibaoka, H. (1981). Changes in microfibril arrangement on the inner surface of the epidermal cell walls in the epicotyl of *Vigna angularis* during cell growth. *Planta (Berlin)*, 151, 385–92.

Tang, R. C. (1973). The microfibrillar orientation in cell wall layers of Virginia pine tracheids. *Wood Science*, 5, 181–6.

Tapp, A. G. (1987). Helicoids in the cell walls of the dwarf French bean (*Phaseolus vulgaris*) and their possible relationship with dietary fibre. B.Sc. thesis, University of Bristol, U.K.

Telfer, W.H., & Smith, D. S. (1970). Aspects of egg formation. In *Insect Ultrastructure*, ed. A. C. Neville, pp. 117–34. Royal Entomological Society of London Symposium, no. 5. Oxford: Blackwell.

Thomas, J., Darvill, A. G., & Albersheim, P. (1983). Characterization of cell wall polysaccharides from suspension-cultured cells of Douglas fir. *Plant Physiology (Bethesda)*, 72, Suppl. 336, p. S.59.

Thompson, D. W. (1961). *On Growth and Form*, ed. J. T. Bonner. Cambridge University Press.

Towe, K. M., & Urbanek, A. (1972). Collagen-like structures in Ordovician graptolite periderm. *Nature (London)*, 237, 443–5.

Tran, T. V. K., Toubart, P., Cousson, A., Darvill,

A. G., Gollin, D. J., Chelf, P., & Albersheim, P. (1985). Manipulation of the morphogenetic pathways of tobacco explants by oligosaccharins. *Nature (London)*, 314, 615–17.

Travis, D. F., & Friberg, V. (1963). The deposition of skeletal structures in the Crustacea. IV. Microradiographic studies of the gastrolith of the crayfish *Orconectes virilis* Hagen. *Journal of Ultrastructural Research*, 8, 48–65.

Trelstad, R. L. (1982). The bilaterally asymmetrical architecture of the submammalian corneal stroma resembles a cholesteric liquid crystal. *Developmental Biology*, 92, 133–4.

Trelstad, R. L., ed. (1984). *The Role of Extracellular Matrix in Development*. New York: Liss Publishing Co.

Trelstad, R. L., & Coulombre, A. J. (1971). Morphogenesis of the collagenous stroma in the chick cornea. *Journal of Cell Biology*, 50, 840–58.

Trelstad, R. L., Hayashi, K., & Gross, J. (1976). Collagen fibrillogenesis: intermediate aggregates and suprafibrillar order. *Proceedings of the National Academy of Sciences, USA*, 76, 4027–31.

Tseng, S. L., Laivins, G. V., & Gray, D. G. (1982). Propanoate ester of (2-hydroxypropyl)-cellulose: a thermotropic cholesteric polymer that reflects visible light at ambient temperatures. *Macromolecules*, 15, 1262–4.

Tyler, S. (1984). Turbellarian platyhelminthes. In *Biology of the Integument. I. Invertebrates*, ed. J. Bereiter-Hahn, A. G. Matoltsy, & K. S. Richards, pp. 112–31. Berlin: Springer.

Valent, B. S., & Albersheim, P. (1974). The structure of plant cell walls. V. On the binding of xyloglucan to cellulose fibres. *Plant Physiology (Bethesda)*, 54, 105–8.

Vian, B. (1978). On the interpretation of twisted patterns in elongating plant cell wall: information obtained with ultracryotomy. *Protoplasma*, 97, 379–85.

Vian, B., Mosiniak, M., Reis, D., & Roland, J.-C. (1982). Dissipative process and experimental retardation of the twisting in the growing plant cell wall. Effect of ethylene-generating agent and colchicine: a morphogenetic revaluation. *Biology of the Cell*, 46, 301–10.

Vian, B., Reis, D., Mosiniak, M., & Roland, J.-C. (1986). The glucuronoxylans and the helicoidal shift in cellulose microfibrils in Linden wood: cytochemistry *in muro* and on isolated molecules. *Protoplasma*, 131, 185–99.

Vian, B., & Roland, J.-C. (1987). The helicoidal cell wall as a time register. *New Phytologist*, 105, 345–58.

Vincent, J. F. V. (1981). Morphology and design of the extensible intersegmental membrane of the adult female locust. *Tissue and Cell*, 13, 831–53.

Vincent, J. F. V. (1982). *Structural Biomaterials*. Basingstoke, U.K.: Macmillan.

Vos, L. De (1974). Etude ultrastructural de la formation et de l'eclosion de gemmules d'*Ephydatia fluviatilis* L. Thése de doctorate, Université Libre de Bruxelles.

Wainwright, S. A. (1988). *Axis and Circumference. The Cylindrical Shape of Plants and Animals*. Cambridge, Mass.: Harvard University Press.

Wainwright, S. A., Biggs, W. D., Currey, J. D., & Gosline, J. M. (1976). *Mechanical Design in Organisms*. London: Arnold.

Walker, L. J., & Crawford, C. S. (1980). Integumental ultrastructure of the desert millipede, *Orthoporus ornatus* (Girard) (Diplopoda: Spirostreptidae). *International Journal of Insect Morphology and Embryology*, 9, 231–49.

Walton, A. G. (1975). Synthetic polypeptide models for collagen structure and function. In *Structure of Fibrous Biopolymers*, ed. E. D. T. Atkins, & A. Keller, pp. 139–50. Colston Papers, Symposium 26 of Colston Research Society, Bristol. London: Butterworth.

Ward, D. V., & Wainwright, S. A. (1972). Locomotory aspects of squid mantle structure. *Journal of Zoology (London)*, 167, 437–49.

Wardrop, A. B., & Harada, H. (1965). The formation and structure of the cell wall in fibres and tracheids. *Journal of Experimental Botany*, 16, 356–71.

Watson, M. R., & Sylvester, N. R. (1956). Studies of invertebrate collagen preparations. *Biochemical Journal*, 71, 578–84.

Weber, F. (1985). Postmoult cuticle growth in a cockroach: *in vitro* deposition of multilamellate and circadian-like layered endocuticle. *Experientia*, 41, 398–400.

Weidinger, M. W., & Ruppel, H. G. (1985). Ca^{2+} requirement for a blue-light-induced chloroplast translocation in *Eremosphaera viridis*. *Protoplasma*, 124, 184–7.

Weis-Fogh, T. (1965). Mechanical properties of insect cuticle. In *Structure and Function of Connective and Skeletal Tissue*, ed. S. Fitton-Jackson, R. D. Harkness, S. M. Partridge, & G. R. Tristram, pp. 373–5. London: Butterworth.

Weis-Fogh, T. (1970). Structure and formation of insect cuticle. In *Insect Ultrastructure*, ed. A. C. Neville, pp. 165–85. Symposia of the Royal Entomological Society of London, no. 5.

Weiss, P. A. (1961). Guiding principles in cell

locomotion and aggregation. *Experimental Cell Research, Supplement, 8,* 260–81.
Weiss, P., & Ferris, W. (1956). The basement lamella of amphibian skin. Its reconstruction after wounding. *Journal of Biophysical and Biochemical Cytology, Supplement 2,* 275–82.
Werbowyj, R. M., & Gray, D. G. (1976). *Liquid Crystalline Structure in Aqueous Hydroxypropylcellulose Solutions.* Pulp and Paper Research Institute of Canada, postgraduate research laboratory report no. 95, Montreal.
Wharton, D. A. (1978). The trichurid egg-shell: evidence in support of the Bouligand hypothesis of helicoidal architecture. *Tissue and Cell, 10,* 647–58.
Wharton, D. A. (1980). Nematode egg-shells. *Parasitology, 81,* 447–63.
Wharton, D. A., & Jenkins, T. (1978). Structure and chemistry of the egg-shell of a nematode (*Trichuris suis*). *Tissue and Cell, 10,* 427–40.
Wiedenmann, G., Lukat, R., & Weber, F. (1986). Cyclic layer deposition in the cockroach endocuticle: a circadian rhythm? *Journal of Insect Physiology, 32,* 1019–27.
Wilkie, K. C. B. (1985). New perspectives on non-cellulosic cell-wall polysaccharides (hemicelluloses and pectic substances) of land plants. In *Biochemistry of Plant Cell Walls,* ed. C. T. Brett & J. R. Hillman, pp. 1–37. Society for Experimental Biology Seminar no. 28 (Glasgow). Cambridge University Press.
Willison, J. H. M., & Abeysekera, R. M. (1988). A liquid crystal containing cellulose in living plant tissues. *Journal of Polymer Science, C: Polymer Letters, 26,* 71–5.
Willison, J. H. M., & Abeysekera, R. M. (1989). Helicoidal arrays of cellulose in quince seed epidermis: evidence for cell wall self-assembly in the plant cell periplasm. *Journal of Applied Polymer Science, Applied Polymer Symposia, 43,* 765–81.
Willison, J. H. M., & Brown, R. M. (1978). A model for the pattern of deposition of microfibrils in the cell wall of *Glaucocystis*. *Planta (Berlin), 141,* 51–8.
Wilson, H. R., & Tollin, P. (1970). Narcissus mosaic virus liquid crystals. *Journal of Ultrastructural Research, 33,* 550–3.
Wolfgang, W. J., Fristrom, D., & Fristrom, J. W. (1986). The pupal cuticle of *Drosophila*: differential ultrastructural immunolocalization of cuticle proteins. *Journal of Cell Biology, 102,* 306–11.
Wolfgang, W. J., Fristrom, D., & Fristrom, J. W. (1987). An assembly zone antigen of the insect cuticle. *Tissue and Cell, 19,* 827–38.
Wolfgang, W. J., & Riddiford, L. M. (1986). Larval cuticular morphogenesis in the tobacco hornworm *Manduca sexta*, and its hormonal regulation. *Developmental Biology, 113,* 305–16.
Wolfgang, W. J., & Riddiford, L. M. (1987). Cuticular mechanics during larval development of tobacco hornworm, *Manduca sexta*. *Journal of Experimental Biology, 128,* 19–33.
Wong, R. (1968). Mechanism of coupling by silanes of epoxies to glass fibers. In *Fundamental Aspects of Fiber Reinforced Plastic Composites,* ed. R. T. Schwarz & H. S. Schwarz, pp. 237–43. New York: Interscience.
Wood, A., & Thorogood, P. (1987). An ultrastructural and morphometric analysis of an *in vivo* contact guidance system. *Development (Cambridge, U.K.), 101,* 363–81.
Woodhead-Galloway, J. (1980). *Collagen: The Anatomy of a Protein.* Institute of Biology Studies in Biology, no. 117. Sevenoaks, U.K.: Edward Arnold.
York, W. S., Darvill, A. G., & Albersheim, P. (1984). Inhibition of 2,4-dichlorophenoxyacetic acid-stimulated elongation of pea stem segments by a xyloglucan oligosaccharide. *Plant Physiology (Bethesda), 75,* 295–7.
Youssef, N. N., & Gardner, E. J. (1975). The compound eye in the opaque-eye phenotype of *Drosophila melanogaster*. *Tissue and Cell, 7,* 655–68.
Yurchenco, P. D. (1990). Assembly of basement membranes. In *Structure, Molecular Biology, and Pathology of Collagen,* ed. R. Fleischmajer, B. R. Olsen, & K. Kühn, pp. 195–213. Annals of the New York Academy of Sciences, no. 580.
Zdarek, J., & Fraenkel, G. (1972). The mechanism of puparium formation in flies. *Journal of Experimental Zoology, 179,* 315–24.
Zelazny, B., & Neville, A. C. (1972). Quantitative studies on fibril orientation in beetle endocuticle. *Journal of Insect Physiology, 18,* 2095–121.

Index

acetoxypropyl cellulose, 132
actin, does not orientate cellulose, 168
aerenchyma cells, 72, 74
age determination by insect cuticle, 172
algae, helicoidal walls of, 66–7
algal cell walls, orthogonal cellulose in, 27–8
annelid cuticle, orthogonal collagen in, 29
annelid eggshell, helicoids in, 49
anti-microtubule drugs, 166
arabinose, 149
arthropod cuticle, helicoids in, 49–53
asymmetrical carbon, optics of, 111, 125–6
auxin, 192; release of oligosaccharins, 119

bark, helicoids in, 73
basal lamina, 31
basement lamella, collagen orientation in, 30–4; helicoids in flatworm, 43, 46; orthogonal collagen in, 32; and tissue evolution, 192
basement membrane, functions of, 32–4
biomimicry of fibrous composites, 124
bird cornea, 36, 38, 114–15, 120, 148, 158
birefringence: anomalous, 112; explained, 105–7; form, 107; intrinsic, 106–7; liquid crystalline collagen, 138; quince mucilage, 143; strain, 106
blue-green algae, 66
blue-phase liquid crystals, 125
Bombyx, helicoidal chorion in, 55–6
bonding, microfibril to matrix, 185
bone, collagen in human, 104; cylindrical orthogonal collagen in human, 38–9; helicoids in human, 62

canal shape in fish eggshells, 61, 118
carbon fibre composite, 93
cell membrane, significance of, 3–4
cells, effect of fibres on, 116
cellulose, 148–9; actin does not orient, 168; fibres, computer patterns of, 177; derivatives, liquid crystalline, 132, 134; helicoid, made in absence of microtubules, 163; and hemicellulose interaction, 144–5, 152–4; microtubules do not align with, 163–4; site of synthesis, 147

cellulose, in sea squirt test, 39–41, 59
cellulose orientation: accuracy in wood of, 174; multinet growth hypothesis, 176–7; plant wall, 65; synthetase packing affects, 168–9, 190–1
chiral twist, helicoids, 187; liquid crystals, 126
chitin, arced pattern grown *in vitro*, 180; in arthropod cuticles, 49–53; crystallinity of, 9–10; experimental change in orientation, 169, 171–2; in fungal spore walls, 70; in nematode eggs, 49; orientation of, 25
chitin orientation, 174
chitin reorientation in insect cuticle: blood-sucking bug, 179; butterfly wings, 180; fly puparium, 179; rhinoceros beetle, 179–80
chitosomes, chitin positioned by, 167–8
cholesteric liquid crystals, 109–10, 125; helical molecules in, 132; long molecules in, 132; molecular asymmetry in, 132; molecular shapes of, 132; xanthan, 188
cholesterics, interference colours in, 131–2
cholesteryl: benzoate, 126; chloride, 190; hexanoate, 190; iodide, 190; nonanoate, 190
chondroitin sulphate, 138, 158
circadian clock, effect on chitin orientation, 26, 169, 171–3
coelacanth, collagen in scales of, 34
collagen, 103–5, 114, 115, 188; bird cornea, 114–15, 148; bone fibre orientation, 64; cellular orientation of, 120; corneal, 35–8; crimped tendon, 26–7; earthworm, 148; fish scale, 34; hierarchical structure of, 140–1; in human bone, 38–9, 62–4; in human discs, 39; liquid crystalline, 138–40; liquid crystals, building units of, 140; in nematode cysts, 47–8; orientation of, 30–4; oriented by fibroblasts, 168; sea cucumber helicoidal, 58; self-assembly, 158; in sponge gemmules, 41, 43; structure of, 87; triple helix, 138–40; type I, 59; types of liquid crystalline, 140

collenchyma cells, 74
colours: interference, 107–8; saturated, 108
composites: convergent evolution of, 192; defined, 9; examples of, 1; long molecular shape of, 185; molecular assembly of, 185; molecular insolubility of, 185; molecular mobility of, 181, 185; principles of, 90–1; proteins in, 86; sizes of components in, 8; stiffness of, 93–5; straight molecules in, 10–11; transitions between types of, 181–4
computer-generated helicoid, 12
conifers, 76
contact guidance by collagen, 119–20
Cook–Gordon model, 92
copolymers, 89
cornea: amphibian, 35; bird, 35–8, 158–9; collagen guides cells in bird, 120; collagen in human, 104–5; fibres influence shape of, 95–6; fish, 35; reptile, 35; sense of ply rotation in vertebrate, 36; shaped by collagen, 115; transparency of bird, 114
cryptobiosis, 53
cuticle: basic structure in crustacean, 50–1; basic structure of insect, 50–3; day layer in insect, 169, 173; dowels in insect, 103; night layers in insect, 169, 173
cyanobacteria, helicoidal mucilage of, 66
cycads, 74
cysts, helicoids in nematode, 46–8; helicoids in protistan, 41

daily layers in insect cuticle, 169, 173; *in vitro* growth, 173
deposition zone, insect cuticular, 147–8
desiccation and helicoidal shells, 100–1
designer composites, 92
developmental biology, 3
dicotyledons, 73
dragonfly thorax design, 103

earthworm, collagen in, 148
egg capsule, fibre orientations in dogfish, 39
eggcase: helicoids in mantis, 54; mantis protein molecules, 188; self-assembly of mantis, 130
eggshell: helicoids in brine shrimp, 53; helicoids in fish, 61–2
eggshell protein molecules: arms of moth, 155–6; central core of moth, 155–6; cloned DNA sequencing of moth, 155; conformation of moth, 155–6; molecular assembly of moth, 155–6; moth, 141; twisted β-pleated sheet of moth, 156
eggshells: functions of fish, 62; functions of moth, 56; helicoids in nematode, 48–9; molecular shape of moth, 189
endosymbiosis of helicoid-secreting ability, 192
evolution, plant cell wall, 192

extensin, 149
extracellular matrix and collagen orientation, 120–1

faults: cofocal parabola, 129–30, 139; in composites, 129–30; in liquid crystals, calculated, 130–1; in liquid crystalline collagen, 139; triple point, 129–30, 139
ferns, helicoids in cell wall of, 71
fibre composites, chemistry of, 85
fibre optics in plants, 115–16
fibre orientation: cell-directed assembly of, 2, 147, 162–4; nine types of, 181–4; parallel examples, 25; primary, 2, 147; secondary, 2–3, 147; self-assembly, 2, 147–8; strategies, 93
fibre strategies in insect cuticle, 101–3
fibre-to-matrix bonding, 87–9
fibreglass principle, 91
fibres: applied aspects of, 6; kinking in worm, 99; and shape, 99
fibroblast traction, collagen oriented by, 168
fibroblasts: collagen secretion by, 36; oriented by collagen, 119–20
fibroin, silk, 26
fish: basement lamella, shingle model of, 175; eggshells, 118, 156; helicoidal eggshell in, 61–2
flight and resilin, 91
flowering plants, helicoids in, 71–3
fluid mosaic model of membrane, 164
fluorescent anti-tubulin, 165
foam, stabilized by helicoid, 96
fossil lobster cuticle, helicoids in, 53–4, 117
fossils: fibres in graptolite, 39; stonewort gyrogonites, 70
fracture planes, insect cuticle, 105
freeze-fracture, 54, 156, 164
fucogalactoxyloglucans: as cell wall helicoids, 119; as hormones, 119
fucose, 149
fungi, helicoids in spore wall of, 70
fungus: in insect cuticle, 118; parasitic, and host cell shape, 121

galactose, 149
gas-liquid chromatography, 152
gemmules, helicoids in sponge, 41, 43
gene expression, 154–5
genetics, moth eggshell protein, 155
geodetics: explained, 14–5, 97, 99; helicoidal, 99–100
glucuronoxylan hemicellulose, 143–5
glucuronoxylans, 149
glycosaminoglycans: constituents of, 86; and liquid crystalline collagen, 138; sea cucumber, 59
graptolite fossils, fibre orientations in, 39
Griffith cracks, 90–1
gyrogonites, 70

helicoid: arced patterns in, 12, 16; computer-generated, 12; corrugated artifact in, 13; critical tilt test for, 17–19; in cyanobacterial mucilage, 145; defined, 15; 3-dimensional representation of a, 15; pitch of, 15; plies of, 11–12; reversal by drug, 158; sea squirt cellulose, 39–41, 59–60; techniques for revealing, 20–1; twist convention, 19; twist reversal, synthetic, 190

helicoids: abundance of 41; in algal cell walls, 66–7; in arthropod cuticle, 49–53; bark, 73; in cell walls of flowering plants, 71–3; consolidation after assembly, 185; constant sense of twist of, 159–61; constraint by previous layer, 186; in cyanobacterial mucilage, 66; discovery in plants, 64–5; duration of assembly, 185; fish eggshell, 61–2; functions of, 95; fungal spore, 70; helical molecules in, 188; history of, 39–40; host–parasite interface, 121; human bone, 62–4; interference colours in, 107–8; isomerism in, 187; liquid crystalline models of, 123; locations in insects, 53; molecular shapes, 187–8; in mollusc shell, 56; monodomain, 186; in moth eggshells, 54–5; neighbouring cells pool secretions, 186–7; peripheral location of, 96; pipe-cleaner shaped molecules, 188; pitch change in, 128–9, 190; plant cell wall, 64–5; polydomain, 186; propellor-shaped molecules, 189; quince mucilage, 73; rope-shaped molecules, 188; self-assembly of, 185; self-perpetuating, 190; shear in, 101; shelled examples of, 100–1; single cell competence, 191; in wood, 74

heliozoan cyst, vesicle secretion in, 41
helix, defined, 14
hemicelluloses: and cellulose interaction, 152–4; constituents, 86; helicoid formers, 149–50; as organizers of helicoids, 149, 152–4; as plant hormone source, 118–19; self-assembly of, 141, 150; site of synthesis, 147; transport route, 166–7; in wood, 151
heparan sulphate, 138
heteropolymer, defined, 148
hierarchical structure, wood, 95
hierarchy, structural, 24
homopolymer, defined, 148
horsetails, helicoids in cell walls of, 71
human: collagen in, 104–5; lumbar discs, collagen in, 103; vertebral discs, fibre orientation in, 39
human bone: collagen in, 104; helicoids in, 62
hyaluronic acid, 114
hydroxypropyl cellulose, liquid crystalline, 132

insect cuticle: orthogonal, 97; perspex model of, 21; screw-carpet model of, 176

insect eggshell, orthogonal protein in, 29
intersegmental membrane, locust, 52

jumping and resilin, 91

keratan sulphate, 158
keratin, a fibrous composite mimic, 7, 124

laminates: Cook–Gordon model of, 92–3; epithelial origin, 6; mechanical failure in, 93
laser Raman spectroscopy, 155
left-handed helicoids, 160–1
lenses: graded refractive index, 113–14; mutant fly, 112–13
lichens, helicoids in cell wall of, 70
light: circularly polarised, 108–10; circularly polarised, in liquid crystals, 131; polarising conventions, 109
lignin, 149
liquid crystals: blue-phase, 125; cholesteric, 109–10, 125; cholesteric quince, 143–4; collagen, 138–40; as model helicoids, 123; molecular sidechains in, 134; natural models, 135; nematic, 124; nematic region of quince, 143; spherulites, 130; stabilization, 134; transitions between types, 181; twist energy of, 126–7; twisted nematic, 125; types of, 124–5; volume changes, 131
lumbar discs, collagen in human, 103

mantis eggcase, orthogonal protein in, 29
mantis oothecin proteins, 136, 138
microfibril-to-matrix bonding, 185
microtubule spring model of cellulose orientation, 165–6
microtubules: and cellulose microfibrils, 163–4; insect muscles attach by, 168
microvilli: and fibre orientations, 174–5; hexagonal lattice of, 175; tetragonal lattice of, 175
modelling bone with sea cucumber collagen, 138
monensin, an anti-helicoid drug, 150–1
monocotyledons, 71–3
mosses, helicoids in cell walls of, 71
moth eggshells: importance of, 55; molecular genetics of, 54–6; morphogenesis of, 154–5; secretion of, 148
mucilage: helicoidal quince, 73; quince, pitch change in, 129
multigene families, 154–5
mycorrhiza, 70

navigation: bird helicoid and, 115; insect helicoid and, 115
nematic liquid crystals, 124
nematodes, helicoids in, 46–9
nutshells, helicoids in, 74

ocelli, insect, 113
oligosaccharins as plant hormones, 118–19, 192

optical activity and helicoid twist, 187
optical rotation: dispersion of, 112; form, 111–12; molecular, 111
orthogonal collagen: not organized by chondroitin sulphate, 158; not organized by keratan sulphate, 158–9
orthogonal fibres, occurrence of, 27
orthogonal plies, patterns in, 11
osteons: bone, 19; collagen orientation in, 38; helicoidal, 62, 64
ovipositor, parallel chitin in insect, 25

papyrus reed, helicoids in, 72
parallel fibre orientation, 102
parasite, fungal, and host cell shape, 121
parenchyma cells, 74
pectin, 149
pericarp cells, 75
periostracum: functions of, 57; helicoidal chitin in, 56
periplasm, plant cell wall, 143, 147
pH can control self-assembly, 161
pH can control stiffness, 161–2
plant cell shape and fibre patterns, 177–9
plant cell wall: bond strengths in, 128; chemistry, 65; compared to insect cuticle, 65; energy hierarchy in, 127–8; functions of, 65; as hormone sources, 118–19; model of, 141; parallel cellulose in, 27; periplasm in, 143
plant cells, types with helicoids, 74–6
plywood, 93; synthetic, 93
polybenzyl glutamate, 126
polyproline II helix, 149
polysaccharides: branched, 86; shape of structural, 11; unbranched, 86
pore canal shape: changes induced in, 117–18; and helicoidal cuticle, 116–17
propellor-shaped molecules, 156
protein export: membrane flow, 4; rough endoplasmic reticulum, 4; signal sequence, 4
protistan cysts, 41
pseudo-orthogonal plies, 97, 102; cellulose orientation, 78; chitin orientation, 77; defined, 21

quince mucilage chemistry, 152

Riemann surface, optics of, 109
reorientation of plant wall cellulose, 176–7; computer predictions of, 177
resilin, 9; birefringence, 106; elasticity of, 52
resilin–chitin composite, 91
reviews, previous, 5
root hair cells, 76
rosettes in plant cell membrane, 164–5
rotifer, helicoids in cuticle of, 49
rubber-like cuticle, insect, 52
rush pith, adaptations of, 97; helicoids in, 71–2

scales of fish, collagen plywood in, 34–5
scar tissue and collagen, 120
sclereid cells, 74, 75

sclerenchyma cells, 74
sclerites, 53
screw axis: cellulose two-fold, 149; pectin three-fold, 149
sea cucumber: helicoids in, 58; orthogonal collagen in, 30
seed epidermis, 75–6
self-assembly: cellulose microfibril 143; contrasted with crystallization, 23–4; crystallites with molecules, 190; defined, 23; effect of pH on, 136; effect of water on, 136: hemicellulose, 141; mantis eggcase proteins, 136; quince mucilage, 142–4; remote control by pH, 161
semi-conductor, 58
serosa cuticle, 51–2
shape, fibres and, 99
shells, spherical, 100–1
signal peptide, moth eggshell proteins, 155
silk, parallel molecules in, 26
skeletal molecules, length of, 1
space-filling network, synthetic, 97
spherulite, liquid crystal, 130
spiral, defined, 15
stellate cells, 74
stereochemical control of helicoids, 160
stiffness: of helicoid, 95; of parallel fibres, 94; of plywood, 95
stone cells, helicoids in, 74
stoneworts, 69–70; helicoidal structure, 69–70
synthetase orients cellulose, 168–9

tadpoles, basement membrane in, 32
template assembly, in plant walls, 24
tendon, parallel chitin in insect, 25; parallel collagen in vertebrate, 26–7; rat tail, 27
tendon cells, insect, 168
terminal-complex rosettes, 164–5
terminal-complexes in plant cell membrane, 164–5
tilt test for helicoids in wood, 81
tracheid cells, conifer, 76
trees, helicoidal parts, 73–4
tunicin, 39

uric acid, in beetle cuticle, 111

Wiener composite, criteria for, 107
wings, fibre orientation in insect, 101
wood: helicoids in, 74; hierarchical structure of, 95; pseudo-orthogonal, 97; structural chemistry of, 151; tracheids, 81
wood fibres, 78, 81; unifying model of, 81; sense of helices in, 78

xanthan, structure of, 188
xylem, cellulose orientation in, 163
xyloglucan, 149
xylose, 149

Young's Modulus, defined, 91